Preface

Quite a large number of scientific and technical problems can be easily tackled with the help of methods developed in Complex Analysis. As a matter of fact, complex analytic functions have enormous potential for practical applications.

The methods developed in complex analysis can be divided into two major categories. In one category, we deal with the problems requiring elementary knowledge of complex numbers; whereas the second one includes the problems requiring a detailed knowledge of the theory of complex analytic functions. The problems connected with the first category are of mechanical vibrating systems, electric circuits, etc., and the problems in the theory of Heat, Fluid mechanics, Electrostatics, dam designs, etc., belong to the second category.

It is not possible to present all the possible applications of a complex variable and complex functions in a single book. The present text is, thus, yet another attempt to present the basic concepts of complex variables and their functions along with some applications in science and engineering.

Suggestions for improvement of the book would be most welcome.

— AUTHORS

Contents

Chapter 1

A Complex Number—Definition and Properties

Art 1.

A number of the form $x + iy$, where x and y are real numbers and i is the imaginary unit, is called a *Complex Number.* For example, $4 + 3i$ and $7 - 5i$ are complex numbers.

If $y = 0$, then $x + iy$ reduces to x which is a real number. Thus, the set of complex numbers contains the real numbers as a subset. If $x = 0$, it reduces to a purely imaginary number iy. The number x is called the *real part* of complex number and y is called its *imaginary part.* It may be noted that y is called imaginary part although it is a real number. Thus, if

$$z = x + iy,$$

then

$$\text{Re} \quad z = x, \quad \text{Im} \quad z = y. \tag{1.1}$$

The complex numbers can be represented as points in a plane called the *complex plane* or *Argand Diagram.* To this end, we choose two perpendicular coordinate axes, the horizontal x-axis and the vertical y-axis, and we take the same unit of length on both the axes. To each complex number $z = x + iy$, there corresponds a point P with cartesian coordinates (x, y) in the complex plane. The horizontal x-axis is called the *real axis*, and the vertical y-axis is known as the *imaginary axis.*

Figure 1 represents the number $2 + 3i$ in the complex plane.

Thus 3i represents a distance 3 units parallel to y-axis. It is, therefore, obvious that 3i is obtained by rotating a distance 3 (representable along x-axis through an angle $\frac{\pi}{2}$ in anti-clockwise sense. Further $-3 = 3i.\ i$ will be obtained by rotating the distance 3 through a further angle $\frac{\pi}{2}$. Thus, we can consider multiplication by i as geometrically equivalent to a rotation along an angle $\frac{\pi}{2}$

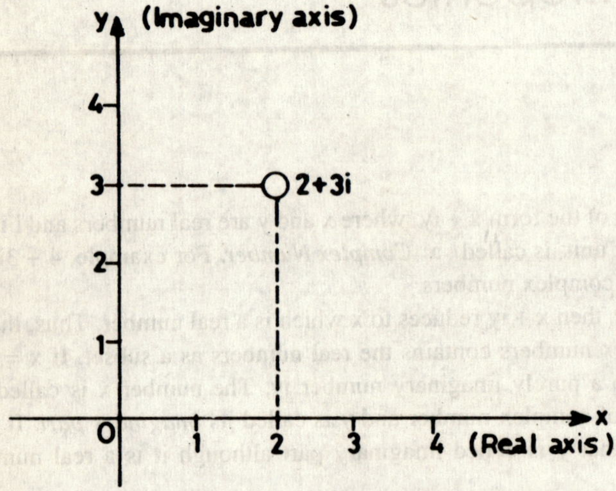

Fig. 1 Argand Diagram.

Let (r, θ) be the polar coordinates of the point P representing the complex number $z = x + iy$ in the complex plane (Fig. 2).

Then, we have

$$x = r\cos\theta, \quad y = r\sin\theta. \tag{1.2}$$

Thus,

$$z = x + iy = r(\cos\theta + i\sin\theta) \tag{1.3}$$

where

$$r = \sqrt{x^2 + y^2}, \tag{1.4}$$

$$\theta = \tan^{-1}\left(\frac{y}{x}\right). \tag{1.5}$$

Equation (1.3) is known as the *Polar Form* or *Trigonometric Form* of a complex number. r is called the *Absolute Value* or *Modulus* of z and is always taken as positive. The angle θ measured positive in counter-

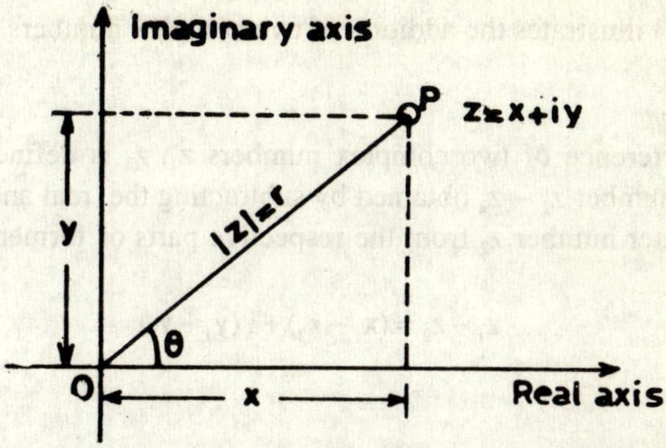

Fig. 2 Polar form of Complex Numbers.

clockwise sense is called the *amplitude* or *argument* of z and is denoted by *amp z* or *arg z*. For $z \neq 0$, the amplitude θ is determined only upto integral multiples of 2π. The value of θ satisfying the inequality.

$$-\pi < \theta \leq \pi \quad (1.6)$$

is called the *principal value* of amplitude of z.

Art 2. EQUALITY AND OPERATIONS WITH COMPLEX NUMBERS

Equality

Two complex numbers

$$z_1 = x_1 + iy_1 \quad \text{and} \quad z_2 = x_2 + iy_2$$

are said to be equal if and only if their real parts are equal and their imaginary parts are equal, *i.e.*,

$$z_1 = z_2 \iff x_1 = x_2 \quad \text{and} \quad y_1 = y_2 \quad (2.1)$$

If z_1 and z_2 are not equal, then we write

$$z_1 \neq z_2.$$

Addition

The sum of two complex numbers z_1, z_2 is defined as the complex number $z_1 + z_2$ obtained by adding their respective real and imaginary parts. That is

$$z_1 + z_2 = (x_1 + x_2) + i(y_1 + y_2) \tag{2.2}$$

Figure 3 illustrates the addition of two complex numbers.

Subtraction

The difference of two complex numbers z_1, z_2 is defined to be the complex number $z_1 - z_2$ obtained by subtracting the real and imaginary parts of later number z_2 from the respective parts of former number z_1. That is

$$z_1 - z_2 = (x_1 - x_2) + i\,(y_1 - y_2) \tag{2.3}$$

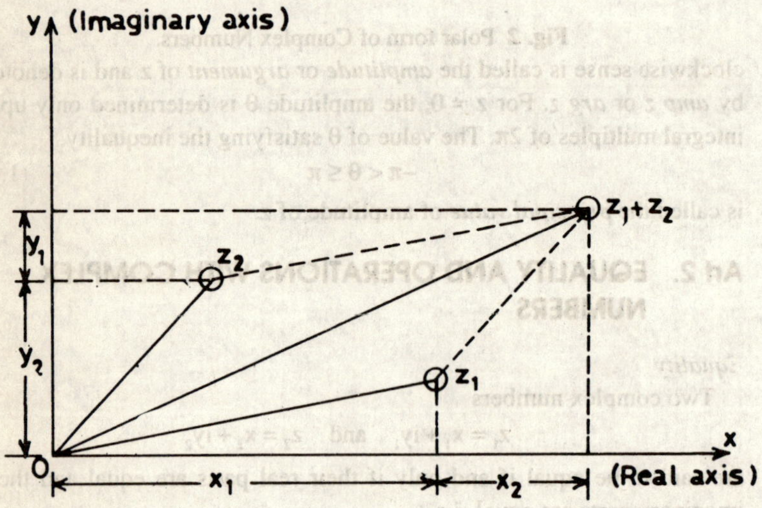

Fig. 3 Addition of Complex Numbers.

In the Figure 4, P denotes the complex number $z_1 - z_2$

It seems that a complex number z can be looked upon as a vector with magnitude $r = \sqrt{x^2 + y^2}$ and direction $\theta = \tan^{-1}\left(\frac{y}{x}\right)$ (angle with x-axis) as at least for addition of two complex numbers the parallelogram law of forces applicable to vectors holds. But it must be remembered that though complex numbers may look like vectors, they are not vectors.

Multiplication

The product of two complex numbers z_1, z_2 is defined as the complex number z_1z_2 obtained by applying the ordinary rules of arithmetic for real numbers, treating the imaginary unit i as a number and replacing its square i^2 by -1. That is

$$z_1z_2 = (x_1 + iy_1)(x_2 + iy_2)$$

$$= (x_1x_2 - y_1y_2) + i(x_1y_2 + x_2y_2). \qquad (2.4)$$

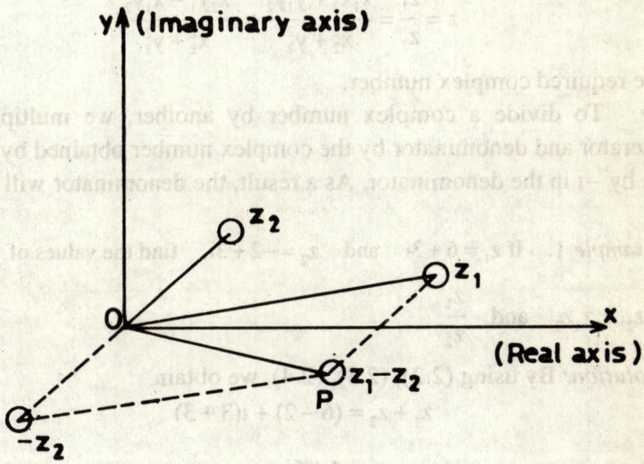

Fig. 4 Subtraction of Complex Numbers.

Division

By the division of a complex number z_1 by another complex number z_2, we mean a complex number z such that

$$z_1 = zz_2. \qquad (2.5)$$

Since z is a complex number, thus

$$z = x + iy.$$

Hence by using (2.4), we get

$$(x_1 + iy_1) = (xx_2 - yy_2) + i(xy_2 + x_2y).$$

Equality of complex numbers (2.1) leads us to write

$$x_1 = xx_2 - yy_2,$$

$$y_1 = xy_2 + x_2y.$$

Hence solving these equations for x and y to obtain

$$x = \frac{x_1x_2 + y_1y_2}{x_2^2 + y_2^2}, \quad y = \frac{x_2y_1 - x_1y_2}{x_2^2 + y_2^2}$$

provided that $x_2^2 + y_2^2$ and so $x_2 + iy_2 = z_2 \neq 0$.

Thus

$$z = \frac{z_1}{z_2} = \frac{x_1x_2 + y_1y_2}{x_2^2 + y_2^2} + i\frac{x_2y_1 - x_1y_2}{x_2^2 + y_2^2}, \tag{2.6}$$

is the required complex number.

Note: To divide a complex number by another, we multiply both numerator and denominator by the complex number obtained by replacing i by –i in the denominator. As a result, the denominator will then be real.

Example 1. If $z_1 = 6 + 3i$ and $z_2 = -2 + 3i$, find the values of $z_1 + z_2$, $z_1 - z_2$, z_1z_2 and $\frac{z_1}{z_2}$

Solution: By using (2.2), (2.3), (2.4), we obtain

$$z_1 + z_2 = (6 - 2) + i(3 + 3)$$

$$= 4 + 6i, \tag{i}$$

$$z_1 - z_2 = (6 + 2) + i(3 - 3)$$

$$= 8. \tag{ii}$$

$$z_1z_2 = (-12 - 9) + i(18 - 6)$$

$$= -21 + 12i, \tag{iii}$$

and

$$\frac{z_1}{z_2} = \frac{6 + 3i}{-2 + 3i}$$

$$= \frac{6 + 3i}{-2 + 3i} \cdot \frac{-2 - 3i}{-2 - 3i}$$

$$= \frac{(-12+9)+i(-18-6)}{4+9}$$

$$= -\frac{3}{13} - \frac{24}{13}i. \qquad \text{(iv)}$$

Art 3. COMPLEX CONJUGATES

If two complex numbers differ only in the sign of their imaginary parts, any of them is said to be the *conjugate* of other.

Thus, if

$$z_1 = x + iy$$

and

$$z_2 = x - iy,$$

then z_1 is called the conjugate of z_2, and z_2 is said to be the conjugate of z_1. From now onwards, we shall denote the conjugate of z by $\bar{z}$ only.

Geometrically, $\bar{z}$ is the reflection of z in the real axis as shown in the Figure 5.

Since

$$z = x + iy$$

and

$$\bar{z} = x - iy,$$

we have

$$z + \bar{z} = 2x, \quad z - z = 2iy, \quad z\bar{z} = x^2 + y^2 \qquad (3.1)$$

Thus

$$x = \text{Re}\, z = \frac{z+\bar{z}}{2}, \qquad (3.2)$$

$$y = \text{Im}\, z = \frac{z-\bar{z}}{2i} \qquad (3.3)$$

Now we shall state and prove the following theorem:

Theorem 1. If z_1 and z_2 are any complex numbers, prove that

$$\overline{z_1 + z_2} = \bar{z}_1 + \bar{z}_2, \qquad (3.4)$$

$$\overline{z_1 - z_2} = \bar{z}_1 - \bar{z}_2, \qquad (3.5)$$

$$\overline{z_1 z_2} = \bar{z}_1 \bar{z}_2, \qquad (3.6)$$

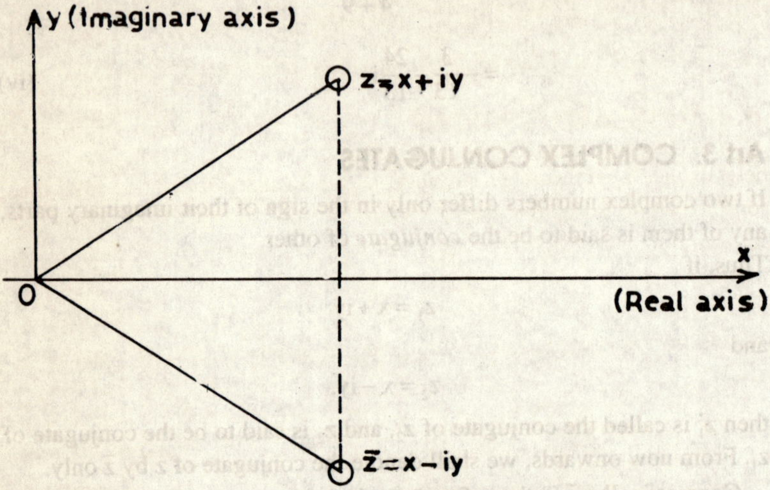

Fig. 5 Complex Conjugates.

$$\overline{\left(\frac{z_1}{z_2}\right)} = \frac{\bar{z}_1}{\bar{z}_2} \tag{3.7}$$

provided for (3.7) z_2 is different from zero.

Proof: Since z_1 and z_2 are complex numbers, we have

$$z_1 = x_1 + iy_1, \quad z_2 = x_2 + iy_2,$$

where x_i and y_i (i = 1, 2) are some real numbers.

By using the definition of complex conjugate, we get

$$\bar{z}_1 = x_1 - iy_1,$$

$$\bar{z}_2 = x_2 - iy_2,$$

and so

$$\bar{z}_1 + \bar{z}_2 = (x_1 + x_2) - i(y_1 + y_2)$$

which by (2.2) is equivalent to $\overline{z_1 + z_2}$.

From the definition of multiplication of complex numbers (2.4), it follows that

$$\bar{z}_1\bar{z}_2 = (x_1x_2 - y_1y_2) - i(x_1y_2 + x_2y_1)$$
$$= \overline{z_1z_2}.$$

The proofs of (3.5) and (3.7) can be given along similar lines.

Art 4. SOME APPLICATIONS

The polar form of complex numbers can be used not only to give the geometrical interpretations of typical arithmetic operations such as addition but also establish the results of multiplication, division of complex numbers without invoking definitions of the same as done above. The polar form of complex number also helps in performing more elaborate calculations which are prohibitive with its cartesian form.

From the last equality in (3.1) and (1.4), it follows that

$$r = |z| = \sqrt{z\bar{z}}. \tag{4.1}$$

Geometrically, $|z|$ is the distance of a point z from the origin 0. Thus, if

$$|z_1| < |z_2|,$$

it is obvious that the point z_2 is farther from the origin than the point z_1. Also, $|z_1 - z_2|$ represents the distance between the points z_1 and z_2.

Example 2. Transform $-1 - i$ to trigonometric form and plot.

Solution: We have,

$$x = -1, \quad y = -1,$$

thus

$$r = \sqrt{(-1)^2 + (-1)^2} = \sqrt{2},$$

$$\tan\theta = \frac{-1}{-1} = 1.$$

Hence, $\theta = 45°$ or $225°$. Since $-1 - i$ lies in the third quadrant, 225° is the correct angle and so

$$\sqrt{2}\,(\cos 225° + i \sin 225°)$$

is the trigonometric form of complex number $-1 - i$, 225° being the principal value of arg z.

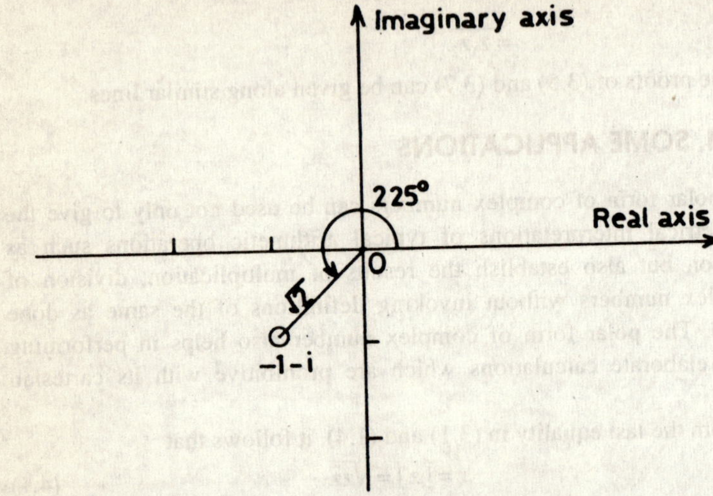

Fig. 6

Now, let

$$z_1 = r_1 (\cos \theta_1 + i \sin \theta_1)$$

and $$z_2 = r_2 (\cos \theta_2 + i \sin \theta_2)$$

be two complex numbers. Then their product z_1z_2 is given by

$$z_1z_2 = [r_1 (\cos \theta_1 + i \sin \theta_1)] \cdot [r_2 (\cos \theta_2 + i \sin \theta_2)]$$

$$= r_1r_2 [(\cos \theta_1 \cos \theta_2 - i \sin \theta_1 \sin \theta_2)$$

$$+ i(\sin \theta_1 \cos \theta_2 + \cos \theta_1 \sin \theta_2)]$$

$$= r_1r_2[\cos (\theta_1 + \theta_2) + i \sin (\theta_1 + \theta_2)]. \quad (4.2)$$

obviously, we have

$$| z_1z_2 | = r_1r_2 = | z_1 | \ | z_2 |, \quad (4.3)$$

$$\arg (z_1z_2) = \theta_1 + \theta_2 = \arg z_1 + \arg z_2 \quad (4.4)$$

Thus in multiplying two complex numbers in trigonometric form, we multiply their absolute values and add their amplitudes.

Geometrically, the distance of z_1z_2 from the origin 0 is equal to the product of the distance of z_1 and z_2 from the origin 0 and arg of z_1z_2 is equal to the sum of args of z_1 and z_2 (Fig. 7).

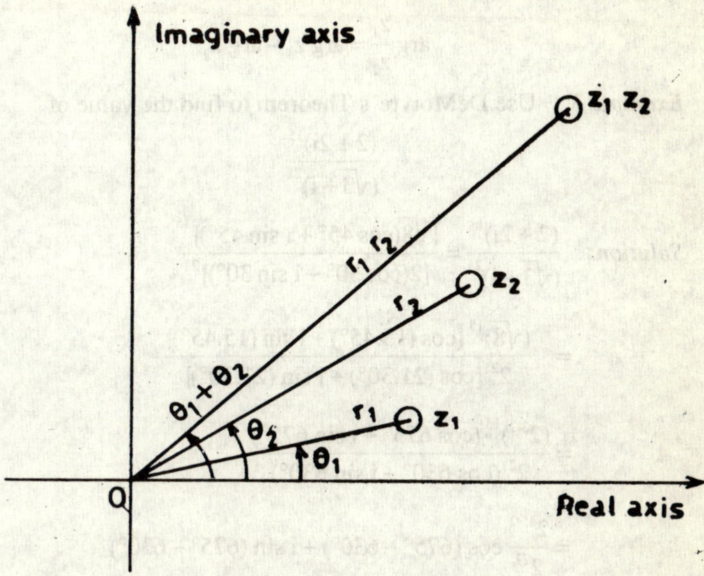

Fig. 7. Multiplication of Complex Numbers.

Generalising the result (4.4) for n complex numbers $z_1, z_2, \ldots\ldots z_n$, we get

$$z_1 z_2 \ldots z_n = r_1 r_2 \ldots r_n\,[\cos(\theta_1+\theta_2+\ldots+\theta_n)+i\sin(\theta_1+\theta_2+\ldots\ldots+\theta_n)]. \tag{4.5}$$

In particular, if all the θ's are same, we have

$$z^n = r^n(\cos n\theta + i\sin n\theta).$$

If r = 1, we obtain the well known form of *DeMoivre's Theorem*

$$(\cos\theta + i\sin\theta)^n = \cos n\theta + i\sin n\theta. \tag{4.6}$$

Similarly, in case of division of complex numbers, we have the following results

$$\frac{z_1}{z_2} = \frac{r_1}{r_2}\,[\cos(\theta_1-\theta_2)+i\sin(\theta_1-\theta_2)], \tag{4.7}$$

$$\left|\frac{z_1}{z_2}\right| = \frac{|z_1|}{|z_2|}, \tag{4.8}$$

and

$$\arg \frac{z_1}{z_2} = \arg z_1 - \arg z_2. \tag{4.9}$$

Example 3. Use DeMoivre's Theorem to find the value of

$$\frac{(2+2i)^{15}}{(\sqrt{3}+i)^{21}}$$

Solution: $\dfrac{(2+2i)^{15}}{(\sqrt{3}+i)^{21}} = \dfrac{[\sqrt{8}(\cos 45° + i \sin 45°)]^{15}}{[2(\cos 30° + i \sin 30°)]^{21}}$

$$= \frac{(\sqrt{8})^{15}[\cos(15.45°) + i \sin(15.45°)]}{2^{21}[\cos(21.30°) + i \sin(21.30°)]}$$

$$= \frac{(2^{3/2})^{15}(\cos 675° + i \sin 675°)}{2^{21}(\cos 630° + i \sin 630°)}$$

$$= \frac{2^{45/2}}{2^{21}} \cos(675° - 630°) + i \sin(675° - 630°)$$

$$= 2^{3/2}(\cos 45° + i \sin 45°)$$

$$= 2\sqrt{2}\left(\frac{1}{\sqrt{2}} + i\frac{1}{\sqrt{2}}\right)$$

$$= 2 + 2i.$$

Definition: A number $w = R(\cos\phi + i \sin\phi)$ is said to be an nth root of the complex number $z = r(\cos\theta + i \sin\theta)$ if $w^n = z$.
That is,
$R^n(\cos\phi + i \sin\phi)^n = r(\cos\theta + i \sin\theta)$
or by DeMoivre's theorem (4.6),
$$R^n(\cos n\phi + i \sin n\phi) = r(\cos\theta + i \sin\theta) \tag{4.10}$$
This will hold good iff $R^n = r$ and $n\phi = \theta + 2k\pi$, where k is some integer.
That is

$$R = n\sqrt{r},$$

$$\phi = \frac{\theta + 2k\pi}{n}$$

For k = 0, 1,....., n – 1, these values of ϕ are all distinct, after this the repetition occurs. Thus, there are n distinct nth roots of a complex number. They have the same absolute value and are given by

$$w_k = n\sqrt{r}\left[\cos\left(\frac{\theta+2k\pi}{n}\right)+i\sin\left(\frac{\theta+2k\pi}{n}\right)\right] \quad (4.11)$$

where k = 0, 1, ..., n – 1.

Example 4. Find the six sixth roots of –729

Solution: We have

$$-729 = 729\,(\cos 180° + i \sin 180°).$$

Therefore,

$$w_k = (729)^{1/6}\left[\cos\left(\frac{180° + k.360°}{6}\right)+i\sin\left(\frac{180° + k.360°}{6}\right)\right]$$

$$= 3\,[\cos(30° + k.60°) + i \sin(30° + k.60°)]$$

where k = 0, 1,....., 5.

Thus

$$w_0 = 3\,(\cos 30° + i \sin 30°) = 3\left(\frac{\sqrt{3}}{2}+i.\frac{1}{2}\right) = \frac{3(\sqrt{3}+i)}{2},$$

$$w_1 = 3(\cos 90° + i \sin 90°) = 3(0 + i) = 3i,$$

$$w_2 = 3(\cos 150° + i \sin 150°) = 3\left(-\frac{\sqrt{3}}{2}+i.\frac{1}{2}\right) = \frac{3(i-\sqrt{3})}{2},$$

$$w_3 = 3(\cos 210° + i \sin 210°) = 3\left(-\frac{\sqrt{3}}{2}+i.-\frac{1}{2}\right)$$

$$= -\frac{3(\sqrt{3}+i)}{2},$$

$$w_4 = 3(\cos 270° + i \sin 270°) = 3(0 + i. - 1) = -3i,$$

$$w_5 = 3(\cos 330° + i \sin 330°) = 3\left(\frac{\sqrt{3}}{2}+i.-\frac{1}{2}\right) = \frac{3(\sqrt{3}-i)}{2},$$ are the six sixth roots of –729.

Now we shall state and prove the following important theorem.

Theorem 2. Prove that the modulus of sum of any two complex numbers is never greater than the sum of their moduli.

Proof: This theorem also known as the *triangle inequality theorem* expresses the fact that the length of any side of a triangle must be equal to or less than the sum of the lengths of the other two sides. That is

$$|z_1+z_2| \le |z_1|+|z_2| \tag{4.12}$$

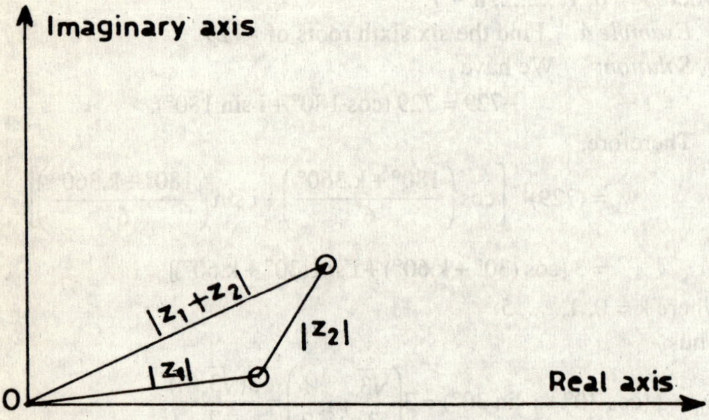

Fig. 8 Triangle Inequality Theorem.

We know that

$$|z|^2 = z\bar{z}$$

Thus

$$\begin{aligned} |z_1+z_2|^2 &= (z_1+z_2)\overline{(z_1+z_2)} \\ &= (z_1+z_2)\ (\bar{z}_1+\bar{z}_2) \\ &= z_1\bar{z}_1+z_2\bar{z}_2+(z_1\bar{z}_2+z_2\bar{z}_1) \\ &= |z_1|^2+|z_2|^2+2\,\text{Re}\,(z_1\bar{z}_2), \end{aligned} \tag{4.13}$$

because

$$\begin{aligned} z_1\bar{z}_2+z_2\bar{z}_1 &= (x_1+iy_1)(x_2-iy_2)+(x_2+iy_2)(x_1-iy_1) \\ &= 2(x_1x_2+y_1y_2) \\ &= 2\,\text{Re}\,(z_1\bar{z}_2). \end{aligned}$$

Since the real part of a complex number is always less than or equal to its absolute value, (4.13) assumes the form

$$|z_1+z_2|^2 \le |z_1|^2+|z_2|^2+2|z_1\bar{z}_2|$$

But $|z_1\bar{z}_2| = |z_1 z_2|$, thus

$$|z_1+z_2|^2 \le |z_1|^2+|z_2|^2+2|z_1 z_2|$$

$$\le (|z_1|+|z_2|)^2$$

Hence

$$|z_1+z_2| \le |z_1|+|z_2|$$

which proves the theorem.

Generalisation of this formula for n complex numbers leads to the following result

$$\left|\sum_{k=1}^{n} z_k\right| \le \sum_{k=1}^{n} |z_k|. \tag{4.14}$$

We now state the following theorems without their proofs.

Theorem 3. The modulus of the difference of two complex numbers is always greater than or equal to the difference of their moduli, i.e. $|z_1-z_2| \ge |z_1|-|z_2|$.

Theorem 4. Prove that

$$|z_1+z_2|^2+|z_1-z_2|^2 = 2(|z_1|^2+|z_2|^2).$$

The proofs of these theorems can be given exactly on the same lines as of theorem 2.

EXERCISE

1. Let $z_1 = 2-3i$, $z_2 = -5+i$. Compute z_1+z_2, z_1-z_2, z_1z_2 and z_1/z_2.
2. Prove that, if a number is equal to its conjugate, it is necessarily real.
3. Prove that any number is equal to the conjugate of its conjugate.
4. Prove that, if the product of two complex numbers is zero, at least one of the numbers must be zero.

Reduce the following expressions to the form a + ib:

5. $(1-2i)(3+2i)^2$.
6. $\dfrac{1+i}{1-i}-\dfrac{1-i}{1+i}$.
7. $\dfrac{(1+i)^3}{(2+i)(1+2i)}$.

8. Find the values of:

$\left|\frac{z}{\bar{z}}\right|$

9. $\operatorname{Im} z^3$

10. $\operatorname{Re}\frac{z}{\bar{z}}$

11. $\left|\frac{1+3i}{3+i}\right|$

12. Verify the triangle inequality for :
$z_1 = 3 - 2i$ and $z_2 = 4 + 2i$.
13. Find all the distinct cube roots of i.
14. Find all distinct fourth roots of -1.
15. Devise a geometrical construction for the quotient of two complex numbers.
16. Prove that the centroid of the triangle whose vertices are z_1, z_2 and z_3 is
$\frac{z_1+z_2+z_3}{3}$.
17. Represent $-1 + i$ in polar form.
18. Using DeMoivre's theorem, express $\cos 4\theta$ and $\sin 4\theta$ in terms of the powers of $\cos\theta$ and $\sin\theta$.
19. Find all solutions of the equation $z^2 - 3z + 3 = i$.
20. Determine all values of $\sqrt[3]{1+i}$.

ANSWERS

1. $z_1+z_2=-3-2i, \quad z_1-z_2=7-4i, \quad z_1z_2=-7+17i,$

$z_1/z_2=\frac{-1+i}{2}.$

5. $-19-22i$.
6. $7i$.
7. $\frac{2}{5}(-1+i)$.
8. 1.
9. $3x^2y-y^3$.

10. $\dfrac{x^2 - y^2}{x^2 + y^2}$.

11. 1.

13. $\dfrac{\pm\sqrt{3} + i}{2}, -i.$

14. $\dfrac{\sqrt{2} \pm i\sqrt{2}}{2}, \dfrac{-\sqrt{2} \pm i\sqrt{2}}{2}$

17. $\sqrt{2}\,(\cos 3\pi/4 + i \sin 3\pi/4)$.

18. $\cos 4\theta = \cos^4\theta - 6\cos^2\theta \sin^2\theta + \sin^4\theta$,

    $\sin 4\theta = 4\,(\cos^3\theta \sin\theta - \cos\theta \sin^3\theta)$.

19. $z = 2 + i,\ 1 - i$.

20. Three values resp. $2^{1/6}(\cos\alpha + i \sin\alpha)$
    where $\alpha = 15°, 135°$ & $255°$

Chapter 2

Functions of a Complex Variable

Art 5. PRELIMINARY DEFINITIONS

Before we start a detailed discussion on analytic functions, we shall explain some mathematical terminologies related with our further studies.

Set of Points. By the term *set of points* in the complex plane, we mean any type of collection of finitely or infinitely many points which may be the points on a line, points in the interior of a circle or solutions of any quadratic equation and so on.

Neighbourhood of a Point. By a *neighbourhood of a point* z_0 in the argand plane we mean any set consisting of all the points which satisfy the inequality of the form

$$|z - z_0| < \varepsilon$$

where ε is an arbitrarily chosen small positive number.

Geometrically, neighbourhood of a point z_0 consists of all points in the interior of circle of radius ε and centre at z_0 but not the points on the circumference of circle.

Open Set. It is a set having the property that its every point has a neighbourhood whose all points belong to the same set. The points in the interior of a circle or square are the examples of such type of sets.

Closed Set. A set is said to be *closed* if its complement, i.e., the set of all points in the complex plane which do not belong to S is open. For example, the points on and inside the circle or square form such type of sets.

Bounded Set. A set is said to be *bounded* if its all points lie within a circle of sufficiently large radius. For example, the points in a square form a bounded set, whereas the points on a straight line do not.

A set which is not bounded is called *unbounded.*

Limit Point. A point z_0 is called a *limit point* of a set if every neighbourhood of z_0 contains at least one point of the set other than z_0.

Connected Set. It is an open set having the property that every pair of its points can be joined by a broken line of finitely many linear segments whose all points belong to the set.

Domain. A set which is open connected is called a *domain.* For example, the interior of a circle is a domain.

Boundary Point. A point of a set is called *boundary point* if its every neighbourhood contains both points belonging to S and not belonging to S. For example, the points on the bounding circles of an annulus are its boundary points.

Region. It is an open connected set with some or all of its boundary points.

Art 6. FUNCTIONS OF A COMPLEX VARIABLE

A complex number which takes only one value in any discussion is called a constant. The numbers considered so far are the examples of constant. On the other hand, a complex number taking different values is said to be a *variable* and is written as

$$z = x + iy$$

where x and y are real variables.

Now let w be a function of z. It means that corresponding to each value of z in a given region R, w has a definite value or set of values. But, if z is given, x and y are known and so any complex function of x and y can be expressed as a function of z. That is

$$w - u(x, y) + iv(x, y) = f(z), \tag{6.1}$$

where u and v are real functions of x and y.

If to each value of z, there corresponds only one value of w then w is said to be a *single valued* function of z. If there are two or more values of w corresponding to one value of z, then w is called a *multi-valued (many-valued)* function of z.

Example 5. If $f(z) = xy + i(x^2 - y^2)$, what is $f(-1 + 2i)$?

Solution: Since

$$x = \text{Re}\, z = -1,$$

$$y = \text{Im}\, z = 2,$$

we have

$$f(-1+2i) = -2 + i(1-4)$$
$$= -2 - 3i.$$

Limit of a Function

A function, defined in a domain D of the complex plane, is said to have the *limit* l as z tends to a point z_0 of D along any path in D, if for every $\varepsilon > 0$, there exists a positive number δ depending on ε and z_0 such that for all points z belonging to D, except z_0,

$$| f(z) - l | < \varepsilon,$$

in the disc $| z - z_0 | < \delta$. It is written as

$$\lim_{z \to z_0} f(z) = l$$

In other words, we may say that l is the limit of f(z) as z approaches z_0 along any path in D if the values of f(z) are as close as desired to l for all z which are sufficiently close to z_0 but are not equal to z_0.

Example 6.

If $f(z) = \dfrac{xy}{x^2 + y^2}$, show that

$$\lim_{x \to 0}\left[\lim_{y \to 0} f(z)\right] = 0 \quad \text{and}$$

$$\lim_{y \to 0}\left[\lim_{x \to 0} f(z)\right] = 0$$

but the $\lim_{z \to 0} f(z)$ does not exist.

Solution: It is clear that

$$\lim_{x \to 0}\left[\lim_{y \to 0} f(z)\right] = \lim_{x \to 0}\left[\lim_{y \to 0} \frac{xy}{x^2 + y^2}\right]$$

$$= \lim_{x \to 0} (0) = 0,$$

$$\lim_{y \to 0}\left[\lim_{x \to 0} f(z)\right] = \lim_{y \to 0}\left[\lim_{x \to 0} \frac{xy}{x^2 + y^2}\right]$$

$$= \lim_{y \to 0} (0) = 0.$$

We also find that $\lim_{z \to 0} f(z)$ along

$y = mx$ is given by

$$\lim_{z \to 0} f(z) = \lim_{z \to 0} \left[\frac{xy}{x^2+y^2}\right] = \lim_{x \to 0} \left[\frac{m}{1+m^2}\right]$$

$$= \frac{m}{1+m^2}$$

which takes different values for different values of m. But by the definition of limit, f(z) should approach the same value along all paths leading to origin, hence $\lim_{z \to 0} f(z)$ does not exist.

Continuity of a Function

A function f(z), defined in a domain D, is said to be *continuous* at a point z_0 of D, if, given any $\varepsilon > 0$, there exists a number δ depending on ε and z_0 such that

$$|f(z) - f(z_0)| < \varepsilon$$

for all points z of D satisfying

$$|z - z_0| < \delta.$$

That is, a function f(z) is continuous at a point z_0 if and only if $f(z_0)$ exists and $\lim_{z \to z_0} f(z) = f(z_0)$.

A function which is continuous at every point of D is said to be *continuous throughout D.*

Example 7. Examine the functions defined as follows for continuity at origin.

$$\text{(i)} \quad f(z) = \frac{\operatorname{Re} z}{|z|}, \quad z \neq 0,$$

$$= 0, \quad z = 0.$$

$$\text{(ii)} \quad f(z) = \frac{(\operatorname{Re} z)^2}{|z|}, \; z \neq 0,$$

$$= 0, \quad z = 0$$

Solution:

$$\text{(i)}\quad f(z) = \frac{x}{\sqrt{x^2+y^2}}$$

Let $z \to 0$ along the radial line $y = mx$, then

$$\lim_{z\to 0} f(z) = \frac{m}{\sqrt{1+m^2}}$$

which will have different values for different values of m. Thus the limit as $z \to 0$ does not exist. The function is thus clearly discontinuous at $z = 0$.

$$\text{(ii)}\quad f(z) = \frac{x^2}{\sqrt{x^2+y^2}}.$$

Let $z \to 0$ along the radial line $y = mx$, then

$$\lim_{z\to 0} f(z) = \lim_{z\to 0}\left[\frac{m^2 x}{\sqrt{1+m^2}}\right] = 0$$

for all m. Even with this, we cannot say with certainity that there is a unique limit as $z \to 0$. For this, we may have to take numerous other possible paths. An alternate method can be the following :

$$\text{For}\quad z \neq 0,$$

$$|f(z)| = \frac{x^2}{\sqrt{x^2+y^2}} \leq |x| \leq |z|$$

which tends to zero as $|z| \to 0$. Hence, the unique limit as $z \to 0$ is zero which is equal to the value of the function at $z = 0$. Thus the function is continuous at $z = 0$.

Differentiability

A function f(z) is *differentiable* at a point $z = z_0$ if

$$\lim_{\Delta z \to 0} \frac{f(z_0+\Delta z) - f(z_0)}{\Delta z} \tag{6.2}$$

exists. We call this limit as the *derivative* of f(z) at the point z_0 and denote it by $f'(z_0)$.

Since the definition of limit implies that z should approach z_0 along any path, differentiability means that (6.2) will always approach the same value of $f'(z_0)$ no matter how Δz approaches zero. Following example clearly explains the importance of this fact:

Example 8. Show that the function

$$f(z) = \bar{z} = x - iy$$

has no derivative.

Solution: We find that

$$\frac{f(z+\Delta z)-f(z)}{\Delta z} = \frac{(x+\Delta x)-i(y+\Delta y)-(x-iy)}{\Delta x + i\Delta y}$$

$$= \frac{\Delta x - i\Delta y}{\Delta x + i\Delta y}.$$

Now, if Δz is real, i.e., $\Delta y = 0$, we have

$$\lim_{\Delta z \to 0} \frac{\Delta x - i\Delta y}{\Delta x + i\Delta y} = \lim_{\Delta x \to 0} \frac{\Delta x}{\Delta x} = 1.$$

On the other hand if Δz is imaginary, i.e., $\Delta x = 0$, we get

$$\lim_{\Delta z \to 0} \frac{\Delta x - i\Delta y}{\Delta x + i\Delta y} = \lim_{\Delta y \to 0} \frac{-i\Delta y}{i\Delta y} = -1.$$

Since the two limits are not same, f(z) has no derivative.

Analytic Function

A function f(z) is *analytic* at a point z_0, if it is defined and possesses a derivative at every point in some neighbourhood of z_0. If a function is differentiable at every point of a domain D of the complex plane, it is said to be *analytic* in D.

We shall now state and prove the necessary and sufficient conditions for a function to be analytic.

Theorem 5.

The *necessary conditions* for a function f(z) = u(x, y) + iv(x, y) to be analytic at the point z are that $\frac{\partial u}{\partial x}, \frac{\partial u}{\partial y}, \frac{\partial v}{\partial x}$ and $\frac{\partial v}{\partial y}$ should exist and satisfy the Cauchy-Riemann differential equations:

$$\frac{\partial u}{\partial x} = \frac{\partial v}{\partial y}, \frac{\partial u}{\partial y} = -\frac{\partial v}{\partial x}. \tag{6.3}$$

Proof: Since f(z) is analytic, f(z) has derivative

$$f'(z) = \lim_{\Delta z \to 0} \frac{f(z+\Delta z)-f(z)}{\Delta z}$$

at every point in some neighbourhood D where Δz may approach zero along any path. If we consider the path of $\Delta z = \Delta x + i\Delta y$ to be real, we then have $\Delta y = 0$, i.e., $\Delta z = \Delta x$ and so

$$f'(z) = \lim_{\Delta x \to 0} \frac{u(x+\Delta x, y) + iv(x+\Delta x, y) - [u(x, y) + iv(x, y)]}{\Delta x}$$

$$= \lim_{\Delta x \to 0} \frac{u(x+\Delta x, y) - u(x, y)}{\Delta x}$$

$$+ i \lim_{\Delta x \to 0} \frac{v(x+\Delta x, y) - v(x, y)}{\Delta x}$$

$$= \frac{\partial u}{\partial x} + i\frac{\partial v}{\partial x}. \tag{6.4}$$

On the other hand, if we consider the path of Δz to be imaginary, we have $\Delta x = 0$, i.e., $\Delta z = i\Delta y$. Thus

$$f'(z) = \lim_{\Delta y \to 0} \frac{u(x, y+\Delta y) + iv(x, y+\Delta y) - [u(x, y) + iv(x, y)]}{i\Delta y}$$

$$= \lim_{\Delta y \to 0} \frac{u(x, y+\Delta y) - u(x, y)}{i\Delta y} + i \lim_{\Delta y \to 0} \frac{v(x, y+\Delta y) - v(x, y)}{i\Delta y}$$

$$= \frac{1}{i} \frac{\partial u}{\partial y} + \frac{\partial v}{\partial y}$$

$$= -i \frac{\partial u}{\partial y} + \frac{\partial v}{\partial y}. \tag{6.5}$$

Since the limits should be independent of paths for the existence of f′(z), equating (6.4) and (6.5) we obtain

$$\frac{\partial u}{\partial x} + i\frac{\partial v}{\partial x} = -i\frac{\partial u}{\partial y} + \frac{\partial v}{\partial y}$$

Now comparison of real and imaginary parts on both sides gives us the Cauchy-Riemann conditions (6.3) which proves the theorem.

Theorem 6. If a function $f(z) = u(x, y) + iv(x, y)$ is such that the four partial derivatives $\frac{\partial u}{\partial x}, \frac{\partial u}{\partial y}, \frac{\partial v}{\partial x}$ and $\frac{\partial v}{\partial y}$ not only exist but are also continuous and satisfy Cauchy-Riemann equations (6.3) then the function f(z) is analytic (*Sufficient Condition*).

Proof: This theorem is the converse of the above theorem and states the sufficient conditions for a function to be analytic. By mean value theorem, we have

$$u(x+\Delta x, y+\Delta y) - u(x, y) = \frac{\partial u}{\partial x}\Delta x + \frac{\partial u}{\partial y}\Delta y, \tag{6.6}$$

$$v(x+\Delta x, y+\Delta y) - v(x, y) = \frac{\partial v}{\partial x}\Delta x + \frac{\partial v}{\partial y}\Delta y. \tag{6.7}$$

Thus

$$f(z+\Delta z) - f(z)$$

$$= u(x+\Delta x, y+\Delta y) + iv(x+\Delta x, y+\Delta y)$$

$$- [u(x, y) + iv(x, y)]$$

$$= u(x+\Delta x, y+\Delta y) - u(x, y)$$

$$+ i\,[v(x+\Delta x, y+\Delta y) - v(x, y)]$$

$$= \left(\frac{\partial u}{\partial x}\right)\Delta x + \left(\frac{\partial u}{\partial y}\right)\Delta y + i\left[\left(\frac{\partial v}{\partial x}\right)\Delta x + \left(\frac{\partial v}{\partial y}\right)\Delta y\right]$$

by using (6.6) and (6.7) in writing the last step. Now employing Cauchy-Riemann conditions (6.3), we obtain

$$f(z+\Delta z) - f(z)$$

$$= \left(\frac{\partial u}{\partial x}\right)\Delta x - \left(\frac{\partial u}{\partial x}\right)\Delta y + i\left[\frac{\partial v}{\partial x}\Delta x + \frac{\partial u}{\partial x}\Delta y\right]$$

$$= \left(\frac{\partial u}{\partial x} + i\frac{\partial u}{\partial x}\right)(\Delta x + i\Delta y)$$

$$= \left(\frac{\partial u}{\partial x} + i\frac{\partial v}{\partial x}\right)\Delta z.$$

Dividing both sides of this equation by Δz, we get

$$\frac{f(z+\Delta z) - f(z)}{\Delta z} = \frac{\partial u}{\partial x} + i\frac{\partial v}{\partial x}.$$

The ratio on left hand side of this equation tends to the limit $\frac{\partial u}{\partial x} + i\frac{\partial v}{\partial x}$ no matter how Δz approaches zero. Thus $f'(z)$ exists and the function $f(z)$ is, therefore, analytic.

Hence the theorem.

Corollary. The derivative of an analytic function w = f(z) is

$$\frac{dw}{dz} = f'(z) = \frac{\partial w}{\partial x} = \frac{1}{i}\frac{\partial w}{\partial y} = u_x + iv_x = \frac{1}{i}(u_y + iv_y).$$

Example 9: Prove that the function u + iv = f(z), where

$$f(z) = \frac{x^3(1+i) - y^3(1-i)}{x^2 + y^2} \quad (z \neq 0),$$

$$f(0) = 0$$

is continuous and that the Cauchy-Riemann conditions are satisfied at the origin, yet f′(0) does not exist.

Solution:

Let us take

$$u = \frac{x^3 - y^3}{x^2 + y^2} \quad \text{and} \quad v = \frac{x^3 + y^3}{x^2 + y^2}.$$

Since f(0) = 0, we have

$$u = 0,\ v = 0.$$

Thus we see that u and v are both rational and finite for all values of z except zero; u, v are, therefore, continuous functions. Hence, f(z) is continuous every where in the domain.

Now the four partial derivatives $\frac{\partial u}{\partial x}, \frac{\partial u}{\partial y}, \frac{\partial v}{\partial x}, \frac{\partial v}{\partial y}$ at the origin are given by

$$\frac{\partial u}{\partial x} = \lim_{x \to 0} \frac{u(x,0) - u(0,0)}{x} = \lim_{x \to 0} \frac{x}{x} = 1,$$

$$\frac{\partial u}{\partial y} = \lim_{y \to 0} \frac{u(0,y) - u(0,0)}{y} = \lim_{y \to 0} \frac{-y}{y} = -1$$

$$\frac{\partial v}{\partial x} = \lim_{x \to 0} \frac{v(x,0) - v(0,0)}{x} = \lim_{x \to 0} \frac{x}{x} = 1,$$

$$\frac{\partial v}{\partial y} = \lim_{y \to 0} \frac{v(0,y) - v(0,0)}{y} = \lim_{y \to 0} \frac{y}{y} = 1.$$

They clearly satisfy the Cauchy-Riemann conditions (6.3).

Further,

$$f'(0) = \lim_{z \to 0} \frac{f(z) - f(0)}{z}$$

$$= \lim_{z \to 0} \frac{x^3 - y^3 + i(x^3 + y^3)}{(x^2 + y^2)(x + iy)}$$

$$= \lim_{z \to 0} \frac{(1+i)(x^3 + iy^3)(x - iy)}{(x^2 + y^2)^2}$$

$$= \lim_{z \to 0} (1+i)\left[\frac{x^4 + y^4 + ixy(y^2 - x^2)}{(x^2 + y^2)^2}\right].$$

When z approaches origin along y = x, we have

$$f'(0) = \frac{1+i}{2}.$$

On the other hand, when z approaches origin along x = 0, we find that

$$f'(0) = 1 + i.$$

Since $f'(0)$ is not unique, $f(z)$ is not analytic at the origin though the Cauchy-Riemann conditions are satisfied there.

Note: In the present problem, the four partial derivatives of first order are not continuous at the origin though it is an essential part of Theorem (6).

Example 10. Is the function $w = e^x \sin y$ analytic?

Solution:

We have

$$u = e^x \sin y,$$

$$v = 0.$$

Thus

$$\frac{\partial u}{\partial x} = e^x \sin y, \frac{\partial u}{\partial y} = e^x \cos y,$$

$$\frac{\partial v}{\partial x} = 0, \frac{\partial v}{\partial y} = 0.$$

Hence, the Cauchy-Riemann conditions become

$$e^x \sin y = 0,$$

$$e^x \cos y = 0.$$

But for no values of x and y, these equations are satisfied. Thus the function is not analytic anywhere in the complex plane.

Note: If a function of complex variable z is purely real or purely imaginary, it will not be analytic.

Theorem 7. If both the real and imaginary parts of an analytic function have continuous second order partial derivatives, then they satisfy Laplace's equation.

$$\frac{\partial^2 \phi}{\partial x^2}+\frac{\partial^2 \phi}{\partial y^2}=0.$$

Proof. Let w = u(x, y) + iv (x, y) be an analytic function of z. Then u and v will satisfy Cauchy-Riemann equations

$$\frac{\partial u}{\partial x}=\frac{\partial v}{\partial y}, \frac{\partial u}{\partial y}=-\frac{\partial v}{\partial x}.$$

Differentiating first equation with respect to x and second with respect to y and adding the two, we obtain

$$\frac{\partial^2 u}{\partial x^2}+\frac{\partial^2 u}{\partial y^2}=0.$$

Similarly, it can be shown that

$$\frac{\partial^2 v}{\partial x^2}+\frac{\partial^2 v}{\partial y^2}=0.$$

Hence the theorem.

Definitions

Any function which possesses continuous partial derivatives of second order and satisfies Laplace's equation is called a *harmonic function.* Two harmonic functions u and v related in the sense that u + iv is an analytic function are known as *conjugate harmonic functions*.

Theorem 8. If w = u(x, y) + iv(x, y) is an analytic function of z, then the family of curves u(x, y) = c_1 and v(x, y) = c_2 in the xy plane intersect at right angles at all their points of intersection.

Proof. For the curve u(x, y) = c_1, we have

$$\frac{\partial u}{\partial x}dx+\frac{\partial u}{\partial y}dy=0,$$

or

$$\frac{dy}{dx}=-\frac{\partial u}{\partial x}\Big/\frac{\partial u}{\partial y}. \tag{6.8}$$

Similarly, for $v(x, y) = c_2$, we obtain

$$\frac{dy}{dx} = -\frac{\partial v}{\partial x} / \frac{\partial v}{\partial y}. \quad (6.9)$$

Since $w = u + iv$ is an analytic function of z, u and v satisfy Cauchy-Riemann conditions. Hence, using (6.3), (6.9) takes the form.

$$\frac{dy}{dx} = \frac{\partial u}{\partial y} / \frac{\partial u}{\partial x}. \quad (6.10)$$

From (6.8) and (6.10) it immediately follows that the curves $u(x, y) = c_1$ and $v(x, y) = c_2$ are orthogonal to each other at all their points of intersection.

Hence the theorem.

EXERCISE

1. Find the real and imaginary part of $f(z) = 2z^3 - 3z$.
2. Express $x^2 + iy^2$ in terms of z and $\bar{z}$.
3. Show that $\lim\limits_{z \to 0} \dfrac{x^2 y}{x^4 + y^2}$ does not exist, even though this function approaches the same limit along every straight line through the origin.
4. Examine the continuity of the function

$$f(z) = \frac{\text{Re}(z^2)}{|z^2|}, \; z \neq 0,$$

$$f(0) = 0.$$

5. Are the following functions analytic ?

   (i) $f(z) = \dfrac{1}{1+z}$,

   (ii) $f(z) = z + \bar{z}$,

   (iii) $f(z) = e^x \cos y$,

   (iv) $f(z) = \arg z$.
6. Find the value of the derivative of $1/z^3$ at 3i.
7. Where are the Cauchy-Riemann conditions satisfied for the function $f(z) = xy^2 + ix^2y$? Where does $f'(z)$ exist ? Where is $f(z)$ analytic ?
8. Show that the function $u = x^3 - 3xy^2 + 3x^2 - 3y^2 + 1$ is harmonic and find the corresponding analytic function.

9. Prove that, if $u = x^2 - y^2, v = -y/(x^2 + y^2)$, both u and v satisfy Laplace's equation but that u + iw is not an analytic function of z.
10. Find an analytic function $w = u + iv = f(z)$ which maps the angular region $0 < \arg z < \frac{\pi}{3}$ onto the region $u < 2$.

ANSWERS

1. $2(x^3 - 3xy^2) - 3x, \quad 2(3x^2y - y^3) - 3y$.
2. $\dfrac{(z^2 + \bar{z}^2)(1 - i)}{4} + \dfrac{z\bar{z}\,(1 + i)}{2}$
4. Discontinuous.
5. (i) Yes $(z \neq -1)$, (ii) No, (iii) No, (iv) No.
6. $-1/27$.
7. Only at origin, only at origin, nowhere.
8. $z^3 + 3z^2 + 1 + ic$ where c is an arbitrary constant.
10. $w = iz^3 + 2$.

Chapter 3

Conformal Mapping

Art 7. CONFORMAL MAPPING

In case of a function of a real variable $y = f(x)$, the plot of y is called the graph of the function x. Thus a single graph with the ordinate y and abscissa x represents the correspondence between the dependent and independent variables y and x.

But in case of a function of complex variable $w = f(z)$

or $u + iv = f(x + iy)$

the situation is entirely different. Here we have four real variables, namely the two dependent variables u and v and the two independent variables x and y. Thus geometrical representation in this case requires a four dimensional space which makes the things more complicated. To overcome this difficulty, we represent the function by means of two planes, namely the z-plane and the w-plane. In the z-plane, we have a plot of y versus x, and in w-plane the plot of v against u. Thus a function $w = f(z)$ is now represented by a correspondence between the points of these two planes. The corresponding points are called *images* of each other, and the point z is said to be *mapped upon* the point (points) w by means of this transformation. As z follows a certain path in the z-plane, the function w traverses a separate path in the w-plane. Thus the function $w = f(z)$ defines a *mapping* or *transformation* of z-plane onto w-plane.

Definitions. If a transformation is such that the angle between the two curves c_1 and c_2 in the z-plane at their point of intersection (x_0, y_0) is equal both in magnitude and sense to the angle at (u_0, v_0), a point corresponding to (x_0, y_0), between the corresponding curves c'_1 and c'_2 in the w-plane, then such a mapping is said to be *conformal*.

A mapping in which the magnitude of angles is preserved only but not necessarily the sense of rotation is termed as *isogonal*. We shall now state and prove the necessary and sufficient conditions for a mapping to be conformal.

Theorem 9. If f(z) is an analytic function of z in a region D of z-plane and $f'(z) \neq 0$ for any point $z \in D$, then the transformation $w = f(z)$ is conformal for all points z belonging to D.

Proof: Let z_0 be the point of intersection of two continuous curves c_1 and c_2 in D. Since f(z) is analytic in D, the curves have definite tangents at z_0 making angles α_1 and α_2 with the real axis. Now let z_1 and z_2 be two points on c_1 and c_2 such that their distance from z_0 is small and same (Fig. 9).

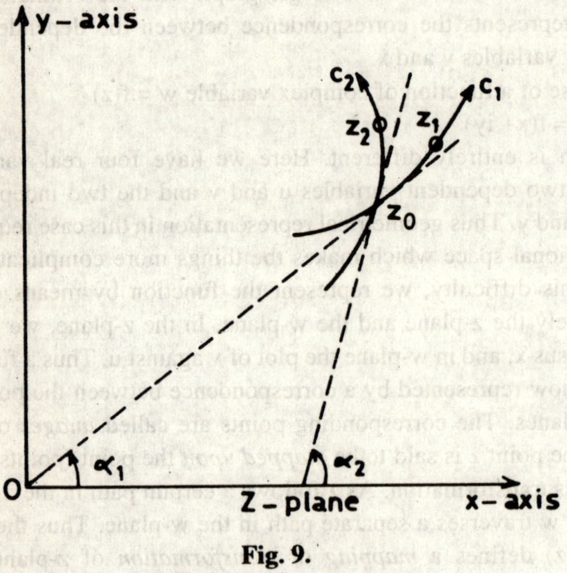

Fig. 9.

Supposing that this distance is r, then we have

$$z_1 - z_0 = r\,e^{i\theta_1}, \qquad (7.1)$$

$$z_2 - z_0 = r\,e^{i\theta_2}, \qquad (7.2)$$

where θ_1 and θ_2 approach to α_1 and α_2, respectively, as r tends to zero.

Now, we assume that the point z_0 is mapped into the point w_0

in the w-plane, and z_1 and z_2 are transformed into w_1 and w_2 describing the curves c'_1 and c'_2 in the w-plane (Fig. 10). We further suppose that

$$w_1 - w_0 = \rho_1 e^{i\phi_1}, \tag{7.3}$$

$$w_2 - w_0 = \rho_2 e^{i\phi_2}, \tag{7.4}$$

where

$$\lim_{z_1 \to z_0} \phi_1 = \beta_1, \quad \lim_{z_1 \to z_0} \phi_2 = \beta_2. \tag{7.5}$$

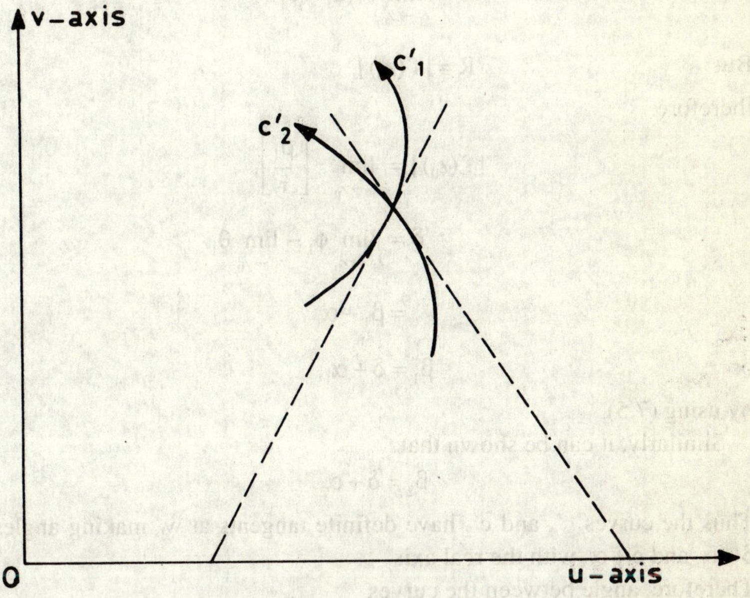

Fig. 10.

Thus we can write the derivative of f(z) at the point z_0 as

$$f'(z_0) = \lim_{z_1 \to z_0} \frac{f(z_1) - f(z_0)}{z_1 - z_0}$$

$$= \lim_{z_1 \to z_0} \frac{w_1 - w_0}{z_1 - z_0} \tag{7.6}$$

Since $f'(z_0) \neq 0$, we let $f'(z_0) = Re^{i\delta}$ and so (7.6), by using (7.1) and (7.3), takes the form

$$Re^{i\delta} = \lim_{z_1 \to z_0} \left[\frac{\rho_1}{r} e^{i(\phi_1 - \theta_1)}\right]$$

By definition of equality of complex numbers, we obtain

$$R = \lim_{z_1 \to z_0} \left[\frac{\rho_1}{r}\right].$$

$$\delta = \lim_{z_1 \to z_0} (\phi_1 - \theta_1).$$

But $$R = | f'(z_0) |,$$

therefore,

$$| f'(z_0) | = \lim_{z_1 \to z_0} \left[\frac{\rho_1}{r}\right],$$

$$\delta = \lim_{z_1 \to z_0} \phi_1 - \lim_{z_1 \to z_0} \theta_1$$

$$= \beta_1 - \alpha_1$$

or $$\beta_1 = \delta + \alpha_1,$$

by using (7.5).

Similarly, it can be shown that

$$\beta_2 = \delta + \alpha_2.$$

Thus the curves c'_1 and c'_2 have definite tangents at w_0 making angles $\delta + \alpha_1$ and $\delta + \alpha_2$ with the real axis.

Therefore, angle between the curves

$$c'_1 \text{ and } c'_2 \text{ at } w_0 = \beta_2 - \beta_1$$

$$= (\delta + \alpha_2) - (\delta + \alpha_1)$$

$$= \alpha_2 - \alpha_1$$

which is same as the angle between the curves c_1 and c_2 at z_0. Furthermore, the sense of measurement of angles in both the cases is also same. Hence, the transformation is conformal which proves the theorem.

The converse of above theorem is :

Theorem 10. If the mapping of z-plane on w-plane is conformal, then the mapping is of the form w = f(z), where f(z) is analytic.

The proof of this theorem can be easily established and is therefore left for the reader.

Theorem 11. If a harmonic function $\phi(x, y)$ is transformed into a function of u, v by means of conformal mapping $w = u + iv = f(z)$, it will then satisfy

$$\frac{\partial^2\phi}{\partial u^2}+\frac{\partial^2\phi}{\partial v^2}=0$$

at all points of the region except possibly at the images of points where $f'(z) = 0$.

Proof: We have

$$\frac{\partial\phi}{\partial x}=\frac{\partial\phi}{\partial u}\cdot\frac{\partial u}{\partial x}+\frac{\partial\phi}{\partial v}\cdot\frac{\partial u}{\partial x}, \tag{7.7}$$

$$\frac{\partial\phi}{\partial y}=\frac{\partial\phi}{\partial u}\cdot\frac{\partial u}{\partial y}+\frac{\partial\phi}{\partial v}\cdot\frac{\partial v}{\partial y}. \tag{7.8}$$

Differenuating (7.7) with respect to x and (7.8) with respect to y and adding the two, we obtain

$$\frac{\partial^2\phi}{\partial x^2}+\frac{\partial^2\phi}{\partial y^2}$$

$$=\frac{\partial\phi}{\partial u}\left(\frac{\partial^2 u}{\partial x^2}+\frac{\partial^2 u}{\partial y^2}\right)+\frac{\partial^2\phi}{\partial u^2}\left[\left(\frac{\partial u}{\partial x}\right)^2+\left(\frac{\partial u}{\partial y}\right)^2\right]$$

$$+\,2\frac{\partial^2\phi}{\partial u\partial v}\left(\frac{\partial u}{\partial x}\frac{\partial v}{\partial x}+\frac{\partial u}{\partial y}\frac{\partial v}{\partial y}\right)$$

$$+\,\frac{\partial\phi}{\partial v}\left(\frac{\partial^2 v}{\partial x^2}+\frac{\partial^2 v}{\partial y^2}\right)+\frac{\partial^2\phi}{\partial v^2}\left[\left(\frac{\partial v}{\partial x}\right)^2+\left(\frac{\partial v}{\partial y}\right)^2\right]. \tag{7.9}$$

Since w is analytic, u and v satisfy Cauchy-Riemann conditions and Laplace's equation. Using these in (7.9), we get

$$\frac{\partial^2\phi}{\partial x^2}+\frac{\partial^2\phi}{\partial y^2}=\frac{\partial^2\phi}{\partial u^2}\left[\left(\frac{\partial u}{\partial x}\right)^2+\left(\frac{\partial v}{\partial x}\right)^2\right]$$

$$+\frac{\partial^2\phi}{\partial v^2}\left[\left(\frac{\partial v}{\partial x}\right)^2+\left(\frac{\partial u}{\partial x}\right)^2\right]$$

$$= \left[\left(\frac{\partial u}{\partial x}\right)^2 + \left(\frac{\partial v}{\partial x}\right)^2\right] \left(\frac{\partial^2 \phi}{\partial u^2} + \frac{\partial^2 \phi}{\partial v^2}\right)$$

$$= | f'(z) |^2 \left(\frac{\partial^2 \phi}{\partial u^2} + \frac{\partial^2 \phi}{\partial v^2}\right). \tag{7.10}$$

Since $\phi(x, y)$ is harmonic, we have

$$\frac{\partial^2 \phi}{\partial x^2} + \frac{\partial^2 \phi}{\partial y^2} = 0.$$

Therefore at points where transformation is conformal, i.e., where $f'(z) \neq 0$, we have

$$\frac{\partial^2 \phi}{\partial u^2} + \frac{\partial^2 \phi}{\partial v^2} = 0$$

which is to be proved.

Art 8. BILINEAR TRANSFORMATION

The transformation of the form

$$w = \frac{az + b}{cz + d} \tag{8.1}$$

where a, b, c, d are complex numbers satisfying the condition $ad - bc \neq 0$ and z, w are complex variables is known as *bilinear* or *linear fractional* or *Mobius transformation.* The expression ad – bc is called the *determinant* of transformation and is non-zero for otherwise (8.1) reduces to a constant showing that the entire z-plane is mapped upon the same point in w-plane. The derivative of (8.1) is found to be

$$\frac{dw}{dz} = \frac{ad - bc}{(cz + d)^2}$$

which is non-zero everywhere and therefore the transformation is conformal except at the point $z = -\frac{d}{c}$ where it is undefined. The mapping (8.1) is one-to one for if $w_1 = w_2$, we have

$$\frac{az_1 + b}{cz_1 + d} = \frac{az_2 + b}{cz_2 + d}$$

or $(ad - bc)(z_1 - z_2) = 0$
implying that $z_1 = z_2$.

We shall now show that the bilinear transformation (8.1) is a combination of the following three transformations:

(i) $w = z + l$

(ii) $w = \lambda z$,

(iii) $w = \dfrac{1}{z}$.

These are respectively known as *translation, rotation* and *inversion* transformations.

i) *TRANSLATION*

In this case, w is obtained by adding a constant vector l to each z. Hence, the transformation is just a translation in the direction defined by arg*l* through a distance equal to $|l|$. The figure in w-plane is the same as the figure in z-plane with a different origin.

ii) *ROTATION*

It is evident in this case that

$$|w| = |\lambda||z|,$$

$$\arg w = \arg \lambda + \arg z.$$

Hence, w is obtained by rotating each z through an angle arg λ and magnifying its length by $|\lambda|$.

The figures in the two planes are similar and similarly situated about their respective origins.

iii) *INVERSION*

In this case, we have

$$w = \frac{1}{z} = \frac{\bar{z}}{z\bar{z}}.$$

Hence, w has the length $\dfrac{1}{|z|}$ and the direction of $\bar{z}$.

Thus the given bilinear transformation (8.1) is a combination of these three transformations for $w = z_3 + \dfrac{a}{c}$, if

$$z_3 = \frac{(b - ad)}{c^2} z_2, \quad z_2 = \frac{1}{z_1}, z_1 = z + \frac{a}{c}.$$

Theorem 12. In the bilinear transformation $w = \dfrac{az + b}{cz + d}$, circle in the z-plane transforms into a circle in w-plane and inverse points transform into inverse points. In the particular case in which the circle in the z-plane transforms into a straight line in the w-plane, the inverse points are transformed into points symmetrical about this line.

Proof: Let the equation of circle in z-plane with inverse points p and q be

$$\left|\frac{z-p}{z-q}\right| = k. \tag{8.2}$$

From the bilinear transformation

$$w = \frac{az+b}{cd+d},$$

we have

$$z = \frac{dw-b}{a-cw}.$$

Putting thi· value in (8.2), we obtain

$$\left|\frac{dw-b-p(a-cw)}{dw-b-q(a-cw)}\right| = k$$

or

$$\left|\frac{w-\dfrac{ap+b}{cp+d}}{w-\dfrac{aq+b}{cq+d}}\right| = k\left|\frac{cq+d}{cp+d}\right|$$

If we assume

$$\frac{ap+b}{cp+d} = P, \quad \frac{aq+b}{cq+d} = Q,$$

$$k\left|\frac{cq+d}{cp+d}\right| = K,$$

then the above equation can be written as

$$\left|\frac{w-P}{w-Q}\right| = K, \tag{8.3}$$

which is the equation of circle with inverse points P and Q.

When $k\left|\dfrac{cq+d}{cp+d}\right| = 1$, then (8.3) represents a straight line which is perpendicular bisector of the line joining p and q. Thus the inverse points of circle in z-plane are transformed into points symmetrical about this line in w-plane.

Hence the theorem.

Example 8. In the mapping $z = \frac{i - w}{i + w}$, show that the positive half of the w-plane given by $v \geq 0$ corresponds to the circle $|z| \leq |$ in the z-plane.

Solution:

We have

$$z = \frac{i - w}{i + w}$$

or

$$x + iy = \frac{i - (u + iv)}{i + (u + iv)}$$

or

$$x + iy = \frac{1 - u^2 - v^2 + 2iu}{u^2 + (1 + v)^2}.$$

Hence

$$x = \frac{1 - u^2 - v^2}{u^2 + (1 + v)^2},$$

$$y = \frac{2u}{u^2 + (1 + v)^2}.$$

Now

$$x^2 + y^2 - 1 = \frac{-4v}{u^2 + (1 + v)^2}. \quad (8.4)$$

Thus when $x^2 + y^2 - 1 = 0$, we have $v = 0$. It means that the circle $x^2 + y^2 = 1$ corresponds to the real line $v = 0$ in w-plane.

Further

$$x^2 + y^2 - 1 < 0$$

implies from (8.4) that $v > 0$. Thus the interior of circle $x^2 + y^2 = 1$ corresponds to the upper half of w-plane as shown in Figure 11′

Theorem 13. The cross ratio of four points z_1, z_2, z_3, z_4 is invariant under a bilinear transformation, *i.e.*

$$\frac{(w_1 - w_2)(w_3 - w_4)}{(w_1 - w_4)(w_3 - w_2)} = \frac{(z_1 - z_2)(z_3 - z_4)}{(z_1 - z_4)(z_3 - z_2)}. \quad (8.5)$$

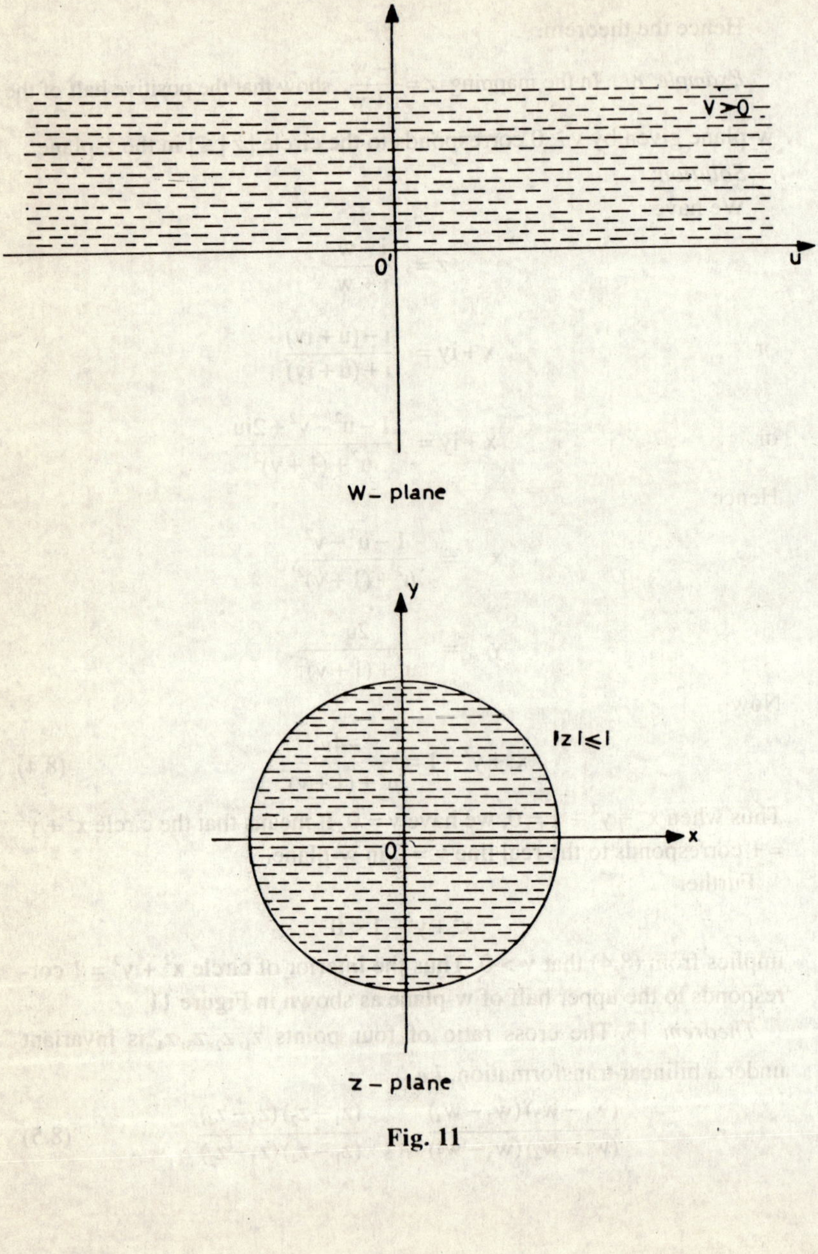

Fig. 11

Proof: We have

$$w_i - w_j = \frac{az_i + b}{cz_i + d} - \frac{az_j + b}{cz_j + d}$$

$$= \frac{(ad - bc)(z_i - z_j)}{(cz_i + d)(cz_j + d)}, \quad (8.6)$$

where i, j = 1, 2, 3, 4.

Thus

$$\frac{(w_1 - w_2)(w_3 - w_4)}{(w_1 - w_4)(w_3 - w_2)}$$

$$= \frac{\frac{(ad - bc)(z_1 - z_2)}{(cz_1 + d)(cz_2 + d)} \cdot \frac{(ad - bc)(z_3 - z_4)}{(cz_3 + d)(cz_4 + d)}}{\frac{(ad - bc)(z_1 - z_4)}{(cz_1 + d)(cz_4 + d)} \cdot \frac{(ad - bc)(z_3 - z_2)}{(cz_3 + d)(cz_2 + d)}}$$

$$= \frac{(z_1 - z_2)(z_3 - z_4)}{(z_1 - z_4)(z_3 - z_2)},$$

which proves the theorem.

Example 9. Find the bilinear transformation which sends the points $z_1 = -1$, $z_2 = 0$, $z_3 = 1$ into the points $w_1 = 0$, $w_2 = i$, $w_3 = 3i$, respectively.

Solution: Employing these values in (8.5) and setting $z_4 = z$, $w_4 = w$, we find that

$$\frac{(0 - i)(3i - w)}{(0 - w)(3i - i)} = \frac{(-1 - 0)(1 - z)}{(-1 - z)(1 - 0)}$$

or

$$\frac{3i - w}{2w} = \frac{1 - z}{1 + z}.$$

We, thus, obtain

$$w = \left(\frac{1 + z}{3 - z}\right) 3i.$$

EXERCISE

1. Does the mapping $w = \bar{z}$ preserve angles in size as well as in sense ?

2. Show that by means of the transformation $w = \left(\frac{z - ic}{z + ic}\right)^2$ the upper half of w-plane may be made to correspond to the interior of a certain semi-circle in the z-plane.

3. Discuss the mapping upon the w-plane of the line $y = mx$ by means of the relation $w = e^z$, m being not equal to zero.
4. Find the points in the z-plane at which the mappings $w = z^5 - 5z$, $w = \sin \pi z$ fail to be conformal.
5. What is the cross ratio of the fourth roots of -1 ?
6. What is the bilinear transformation which sends the points $z = 0$. -1, ∞ into the points $w = -1, -2 - i, i$ respectively? What is the image of the circle $|z| = 1$ under this transformation ?
7. What is the most general bilinear transformation which maps the upper half of the z-plane onto the lower half of the w-plane ?
8. Find a transformation which will map a 45° sector of the unit circle in the z-plane onto the upper half of the w-plane.
9. Show that, if $|c| = |d|$, then the transformation $w = \dfrac{az+b}{cz+d}$ maps the unit circle in the z-plane into a straight line in the w-plane.
10. Prove that four points z_1, z_2, z_3, z_4 lie on a circle if and only if their cross ratio is real.
11. Show that the relation $w = \dfrac{iz+2}{4z+i}$ transforms the real-axis in the z-plane into a circle in the w-plane. Find the centre and radius of the circle and the point in the z-plane which is mapped on the centre of the circle.
12. By the transformation $w = z^2$, show that the circle $|z - a| = c$, (where a and c are real) in the z-plane corresponds to the limacon in the w-plane.
13. Find the images of the following regions, under the mapping $w = e^z$:

    (i) $-3 \le x \le 3, \quad -\pi \le y \le -\dfrac{\pi}{2}$

    (ii) $1 < x < 2, \quad 0 < y < 1.$
14. Find the image of the region $0 < x < 2\pi, \quad 1 < y < 2$ under the mapping $w = \sin z$.

ANSWERS

1. No, the size is preserved, but the sense is reversed.
3. Spiral.

4. $z = \pm 1, \pm i; \quad z = \frac{(2n+1)}{2}, \quad n = 0, \pm 1, \pm 2, \ldots$

5. -1.

6. $w = \frac{iz-2}{z+2}; \quad 3(u^2+v^2)+8u+2v+3=0.$

7. $w = \frac{az+b}{cz+d}$, where a, b, c, d are all real and $ad - bc < 0$.

8. $w = -\left(\frac{z^4-1}{z^4+1}\right)^2.$

11. $w = -\frac{7}{8}i, \frac{9}{8}, z = \frac{i}{4}.$

13 (i) $e^{-3} \le |w| \le e^3, \quad -\pi \le \arg w \le -\frac{\pi}{2}$

(ii) $e < |w| < e^2, \quad 0 < \arg w < 1.$

14. Elliptic annulus bounded by ellipses

$\frac{u^2}{\cosh^2 1} + \frac{v^2}{\sinh^2 1} = 1$ & $\frac{u^2}{\cosh^2 2} + \frac{v^2}{\sinh^2 2} = 1$ and cut along the positive imaginary axis.

Chapter 4

Schwarz—Christoffel Transformation

Art 9. A SPECIAL CONFORMAL TRANSFORMATION

One of the most applicable transformations is the Schwarz-Christoffel transformation. It maps the interior of a polygon in the w-plane in such a manner that the sides of the polygon in the w-plane are mapped into the real axis of z-plane.

Theorem 14. Let w_1, w_2, ... w_n, be the n vertices of a closed polygon in w-plane and the corresponding interior angles be α_1, α_2 ....α_n, respectively, so that

$$\alpha_1 + \alpha_2 + \ldots\ldots + \alpha_n = (n-2)\pi.$$

Let the points w_1, w_2, w_3,........w_n be mapped into the points $x_1, x_2, x_3, \ldots\ldots, x_n$ arranged in ascending order on the real axis of z-plane, then the Schwarz-Christoffel transformation is given by the differential equation.

$$dw/dz = K(z-x_1)^{\frac{\alpha_1}{\pi}-1} \quad (z-x_2)^{\frac{\alpha_2}{\pi}-1} \quad \ldots\ldots(z-x_n)^{\frac{\alpha_n}{\pi}-1} \qquad (9.1)$$

Here K is a constant which may be a complex number.

Proof: Four aspects of this theorem have to be examined to prove it, *viz.*,

i) as to what happens when z increases from x_1 to x_2;
ii) what happens when z passes through x_2;
iii) what region of the w-plane corresponds to the upper and lower halves of the z-plane; and
iv) what happens when z goes from $-\infty$ to $+\infty$ along the real axis.

Let us examine these aspects one by one and see what it establishes. It is clear from (9.1) that due to the vanishing of $z - x_1$ at $z = x_1$, it follows that dw/dz is either zero or infinite according as $\alpha_1 > \pi$ or $< \pi$. The points $x_1, x_2, \ldots\ldots x_n$ on the real axis are thus to be avoided, which can be done by enclosing them with semi-circles having these points as centres, of small radius and situated in the upper half of the z-plane. Let the semi-circle with centre at x_1 cut the real axis in points a_1 and a_2 (Fig.12), we can take z to describe the real axis in the positive direction even at $z = x_1$, (so that dz is positive even at $z = x_1$). Similar argument can be given for $z = x_2, x_3, \ldots\ldots, x_n$ also. It stands therefore established that when z goes from $z = x_1$ to $z = x_2$ in the z-plane the corresponding points in the w-plane describe a straight line. (See Fig. 12 for corresponding geometries in the two planes).

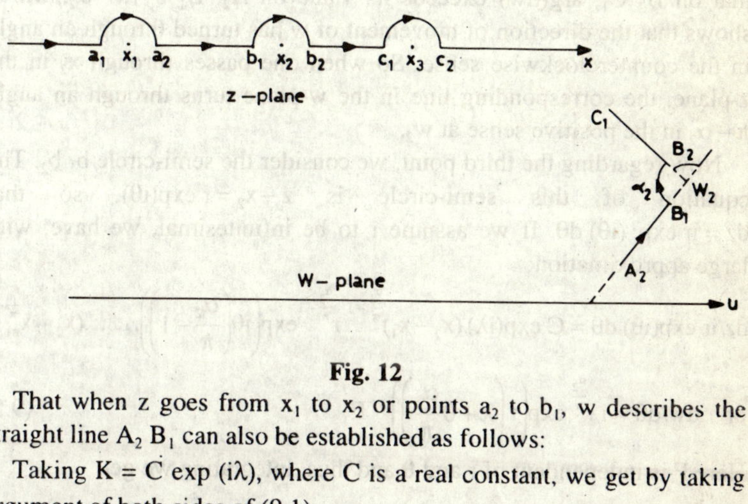

Fig. 12

That when z goes from x_1 to x_2 or points a_2 to b_1, w describes the straight line A_2 B_1 can also be established as follows:

Taking $K = C \exp(i\lambda)$, where C is a real constant, we get by taking argument of both sides of (9.1)

$$\arg(dw) - \arg(dz) = \lambda + \left(\frac{\alpha_1}{\pi} - 1\right)\arg(z - x_1) + \left(\frac{\alpha_2}{\pi} - 1\right)\arg(z - x_2) + \ldots\ldots + \left(\frac{\alpha_n}{\pi} - 1\right)\arg(z - x_n). \quad (9.2)$$

From (9.2) it is clear that as z moves from a_2 to b_1, arg(dz) = 0, arg $(z - x_1) = 0$ since $z - x_1$ is real and positive; $\arg(z - x_i) = 0$ (i = 2, 3, ...n), since $z - x_i$, i = 2, 3,.....n are all real and negative. Thus

$$\arg(dw) = \lambda + (\alpha_2 - \pi) + (\alpha_3 - \pi) + (\alpha_4 - \pi) + \ldots\ldots + (\alpha_n - \pi). \tag{9.3}$$

This shows that arg (dw) is constant as z moves a_2 to b_1, or in other words w describes a straight line $A_2 B_1$.

Now we examine as to what happens when z passes through x_2.

The semi-circle which isolates the point x_2 on the real axis cuts the real axis at points b_1 and b_2, while that for x_3 cuts it in the points c_1 and c_2. When z goes from x_2 to x_3 on the real axis the corresponding points in the w-plane describe a straight line as already established. Now as z increases from b_2 to c_1, we get from (9.2):

$$\arg(dw) = \lambda + (\alpha_3 - \pi) + (\alpha_4 - \pi) + \ldots\ldots + (\alpha_n - \pi). \tag{9.4}$$

From (9.4) we find that w describes the straight line $B_2 C_1$ and also that on $B_2 C_1$, arg(dw) exceeds its value on A_2, B_1 by $\pi - \alpha_2$, which shows that the direction of movement of w has turned through an angle in the counterclockwise sense. So when one passes through x_2 in the z-plane, the corresponding line in the w-plane turns through an angle $\pi - \alpha_2$ in the positive sense at w_2.

Now regarding the third point, we consider the semi-circle b_1 b_2. The equation of this semi-circle is $z - x_2 = r\exp(i\theta)$, so that $dz = ir\exp.(i\theta)\, d\theta$. If we assume r to be infinitesimal, we have, with large approximation:

$$dz/ir\exp(i\theta)\, d\theta = C\exp(i\lambda)(x_2 - x_1)^{\frac{\alpha_1}{\pi}-1}\, r^{\frac{\alpha_2}{\pi}-1} \exp\left(i\theta\left(\frac{\alpha_2}{\pi} - 1\right)\right)\ldots\ldots(x_2 - x_n)^{\frac{\alpha_n}{\pi}-1}$$

$$\text{or} \quad dw/d\theta = i\, r^{\frac{\alpha_2}{\pi}} \exp\left(i\left(\lambda + \theta\frac{\alpha_2}{\pi}\right)\right) F \tag{9.5}$$

Here F is independent of r and θ and thus, integrating we get:

$$w = w_1 + \frac{\pi}{\alpha_2} r^{\frac{\alpha_2}{\pi}} \exp\left(i\left(\lambda + \theta\frac{\alpha_2}{\pi}\right)\right) F \tag{9.6}$$

w_1 is a constant of integration which tends to zero as r tends to zero. This w_1, therefore, corresponds to x_1. This transformation thus makes w to describe a polygon with vertices corresponding to points x_1, x_2, x_3, ...x_n and internal angles $\alpha_1, \alpha_2, \ldots\ldots \alpha_n$.

From (9.6), we have:

$$\arg(w - w_1) = \lambda + \theta \frac{\alpha_2}{\pi} + \arg F. \tag{9.7}$$

Thus with z describing the semi-circle, θ decreases from π to 0, so that arg (w − w_1) decreases by α_2 and consequently w describes a circular arc with centre x_2, situated inside the polygon. It well establishes that the points within the polygon, or the interior of the polygon, maps on to the upper half of the z-plane. The exterior region will obviously map on to the lower half of the z-plane.

Lastly, we have to examine as to what happens when z progresses from −∞ to +∞ along the real axis in the z-plane. We consider a triangle, which is the simplest form of a polygon. We can take the real axis in the z-plane accordingly indented at three points x_1, x_2, x_3 and a large semi-circle having its centre at origin. (Fig.13).

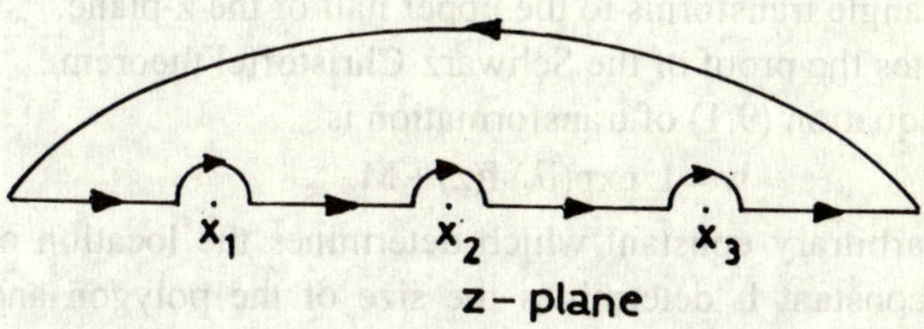

Fig. 13

When z goes from x_1 to x_2 and x_2 to x_3 we get corresponding sides AB and BC of the triangle indented at A, B, C (Fig. 14)

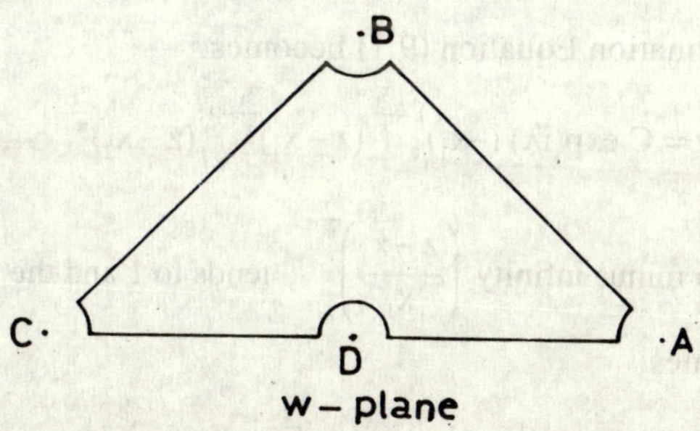

Fig. 14

Let the big semi-circle have radius R, then its equation is $z = R\exp(i\theta)$. For large R we can, with sufficient approximation replace $z - x_1$, $z - x_2$ and $z - x_3$ by $R\exp(i\theta)$. Equation (9.1) then yields:

$$dw/iR\exp(i\theta)\,d\theta = C\exp(i\lambda)\,[R\exp(i\theta)]^{\frac{\alpha_1+\alpha_2+\alpha_3}{\pi} - 3}.$$

Now since $\alpha_1 + \alpha_2 + \alpha_3 = \pi$, we get:

$$dw/d\theta = iC\exp(i(\lambda - \theta)), \text{ giving on integration } w = w_D - \frac{C}{R}\exp i.\,(\lambda - \theta)$$

Here w_D is a constant to which w tends when R tends to infinity.

It is clear from the above that argument of $w - w_D$ is $\pi + \lambda - \theta$. Thus, when z describes the large semi-circle, θ goes from 0 to π and arg $(w - w_D)$ goes from $\pi + \lambda$ to λ. In other words, w describes the semi-circle of small radius C/R about the point D (Fig. 14) which tends to zero as R tends to infinity. It is hence concluded that the region within the indented triangle transforms to the upper half of the z-plane.

This completes the proof of the Schwarz-Christoffel theorem.

Integral of Equation (9.1) of transformation is:

$$w = L\exp(i\lambda)\,f(z) + M. \qquad (9.8)$$

Here M is an arbitrary constant which determines the location of the polygon. The constant L determines the size of the polygon and the constant λ its orientation.

Let us now analyse the situation when a vertex of the polygon corresponds to a point at infinity on the real axis of the z-plane. Let x tend to minus infinity. In this case we can choose K as $C\exp(i\lambda)\,(-x_1)^{1-\frac{\alpha_1}{\pi}}$, so that the transformation Equation (9.1) becomes:

$$dw/dz = C\exp(i\lambda)\,(-x_1)^{1-\frac{\alpha_1}{\pi}}\,(z - x_1)^{\frac{\alpha_1}{\pi}-1}\,(z - x_2)^{\frac{\alpha_2}{\pi}-1}\,\ldots.. \qquad (9.9)$$

When x tends to minus infinity $\left(\dfrac{z - x_1}{x_1}\right)^{\frac{\alpha_1}{\pi}-1}$ tends to 1 and the transformation (9.9) becomes:

$$dw/dz = C\exp(i\lambda)\,(z - x_2)^{\frac{\alpha_2}{\pi}-1}\,(z - x_3)^{\frac{\alpha_3}{\pi}-1}\,\ldots.. \qquad (9.10)$$

which shows that the factor corresponding to $x_1 = -\infty$ is omitted from the equation of transformation or the angle α_1 does not appear.

Corollary: If the points $x_1, x_2, x_3, \ldots\ldots, x_n$, lie on the circumference of unit circle $|z| \leq 1$, rather than on real axis in the z-plane, then the transformation (9.1) maps inside of the polygon on to the inside of the unit circle.

Proof: We have

$$x_1 = \exp(iA), \quad x_2 = \exp(iB), \ldots.., \quad z = \exp(i\eta)$$

and thus,

$dz/d\eta = i \exp(i\eta)$ on the circumference of the circle.

Since $\alpha_1 + \alpha_2 + \ldots\ldots + \alpha_n = (n-2)\pi$, (9.1) gives

$$dw/d\eta = dw/dz \,.\, dz/d\eta = dw/dz \,.\, i \exp(i\eta)$$

$$= K \left(e^{i(\eta - A)}\right)^{\frac{\alpha_1}{\pi}-1} \left(e^{i(\eta - B)}\right)^{\frac{\alpha_2}{\pi}-1} \ldots\ldots\ldots$$

or $$dw/dz = K_1 \left[\sin\frac{1}{2}(\eta - A)\right]^{\frac{\alpha_1}{\pi}-1} \left[\sin\frac{1}{2}(\eta - B)\right]^{\frac{\alpha_2}{\pi}-1} \ldots\ldots \tag{9.11}$$

where

$$k = ik\left(e^{\frac{iA}{2}}\right)^{\frac{\alpha_1}{\pi}-1} \left(e^{\frac{iB}{2}}\right)^{\frac{\alpha_2}{\pi}-1} \ldots\ldots. \tag{9.12}$$

When z advances from A to B, the arg(dw/dz) remains constant and consequently w describes the straight line $A_2\ B_2$ in the w-plane. By going round the point x_2 on the semicircle $b_1\ b_2$ inside the unit circle (Fig. 15), $\arg(z - x_2)$ increases by $\pi - \alpha_2$ and this amounts to turning

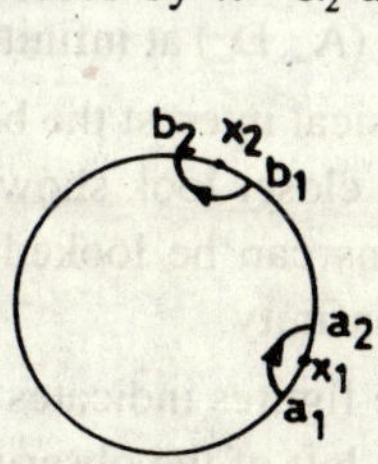

Fig. 15

through the corner X_2 in the w-plane. (See Fig.12 also)

The corollary is thus proved.

Extended Definition of A Polygon: The popular idea of a polygon is, in fact, a closed figure like a triangle, rectangle, pentagon and so on. The interior and exterior of a closed polygon are well recognized regions. However, how about rectilinear configurations which may at first sight not appear to be polygons? Through a closer look, however, these rectilinear configurations can also be looked upon as polygon, one or more of whose vertices are at infinity. The concept of exterior and interior of a polygon has thus to be defined more precisely. **That Region which lies to the left of an observer traversing the boundary of the polygon in a certain sense will be regarded as its interior, the region on the other side of the boundary being regarded as exterior of the polygon.**

The usual convention followed in this regard is that **The Interior Region is the one which lies to the left of an observer traversing the boundary in the counter clockwise direction with his head held towards us, the other side being the exterior region.**

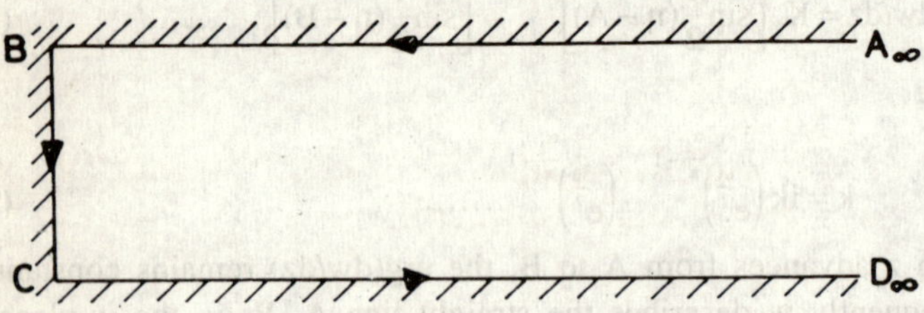

Fig. (i) - A rectangle with two vertices A_∞ and D_∞ at infinity, or alternatively a triangle with vertex (A_∞, D_∞) at infinity.

In most problems of physical interest the boundaries, at the face of it, do not look polygonal but a closer look shows them as such. For example, following configurations can be looked upon as polygons; suffix infinity, denoting points at infinity.

The hedging in the above figures indicates the exterior region and the interior region (lying on the left of the observer traversing the boundary in sense denoted by arrows) is on the other side of the boundary.

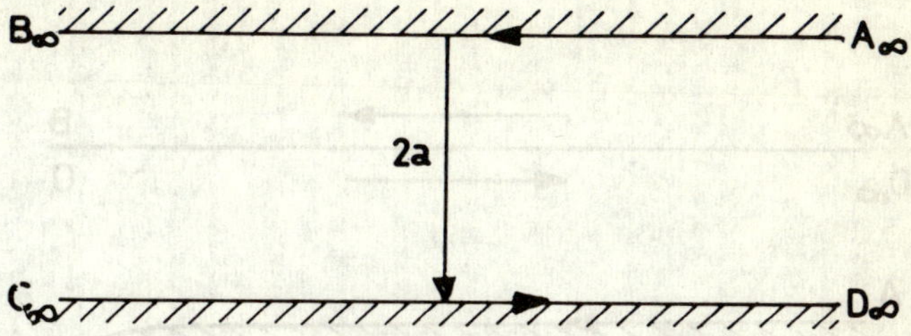

Fig. (ii) - A rectangle with all the vertices at infinity.

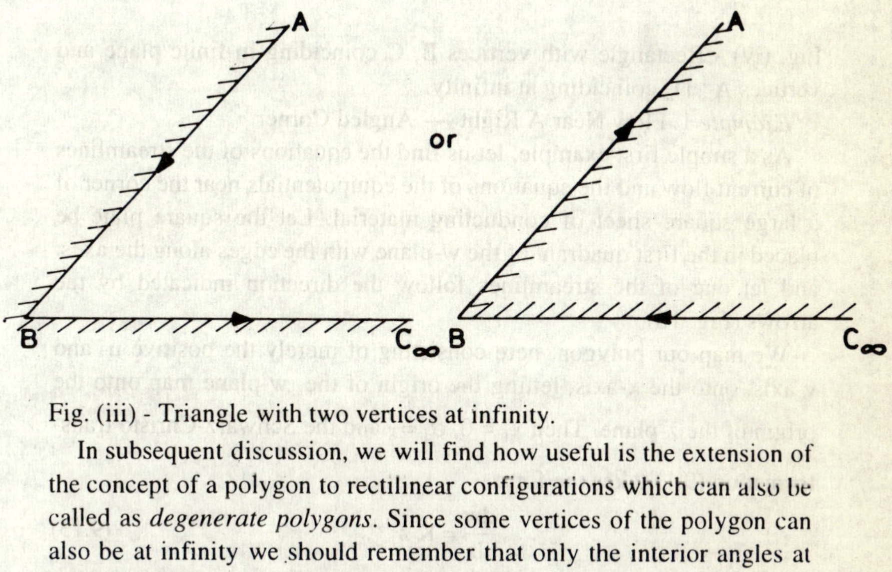

Fig. (iii) - Triangle with two vertices at infinity.

In subsequent discussion, we will find how useful is the extension of the concept of a polygon to rectilinear configurations which can also be called as *degenerate polygons*. Since some vertices of the polygon can also be at infinity we should remember that only the interior angles at vertices which map onto points in finite z-plane together with only these image points will enter the Schwarz-Christoffel transformation. Also if the polygon is symmetrical about a certain point it will be convenient if we map this point onto origin in w-plane.

We will now demonstrate the use of Schwarz-Christoffel transformation in solving some problems of physical interest.

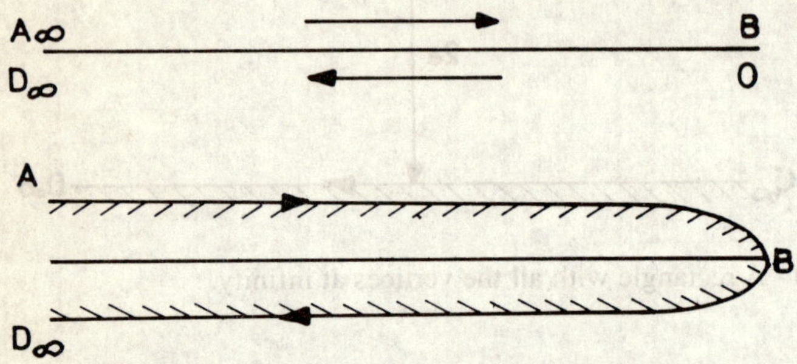

Fig. (iv) - Rectangle with vertices B, C coinciding in finite plane and vertices A_∞, D_∞ coinciding at infinity.

Example 1. Flow Near A Right — Angled Corner

As a simple first example, let us find the equations of the streamlines of current flow and the equations of the equipotentials near the corner of a large square sheet of conducting material. Let the square plate be placed in the first quadrant of the w-plane with the edges along the axes, and let one of the streamlines follow the direction indicated by the arrows (Fig. 16).

We map our polygon, here consisting of merely the positive u- and v-axis, onto the x-axis, letting the origin of the w-plane map onto the origin of the z-plane. Then $x_1 = 0$, $\alpha_1 = \frac{\pi}{2}$ and the Schwarz-Christo transformation (9.1) takes the form.

$$\frac{dw}{dz} = K z^{-1/2}, \tag{9.13}$$

whence $$w = 2 K z^{1/2} + K'.$$

Since z = 0 when w = 0, the constant of integration K′ will be zero, and we have

$$w = 2 K z^{1/2}$$

or $$z = \frac{w^2}{4K^2}. \tag{9.14}$$

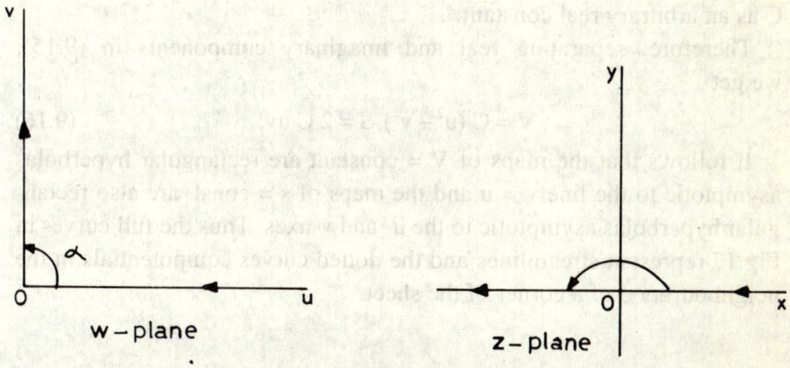

Fig. 16

Since the boundary streamline in the w-plane maps into the x-axis streamline from right to left, we may put into the z-plane a parallel flow of current such that the x-axis is one of the streamlines. Hence, the mapping from the z-plane to the final F-plane may be brought about by the simple transformation F = k z, where k is a real constant depending upon the system of units, or scale, used: this transformation will certainly change the net of straight lines in the z-plane into a similar net in the F-plane. Then (9.14) becomes

$$F = \frac{k}{4k^2} w^2 = C\, w^2, \tag{9.15}$$

where C is a constant. If we now think of the relation (9.15) as mapping from the F-plane to the w-plane, we see that the straight horizontal streamlines of our parallel current flow and the vertical equipotentials orthogonal to the streamlines are transformed into the corresponding streamlines and equipotentials in the first quadrant of the w-plane, *i.e.* into the desired curves.

The constant of proportionality C in (9.15) may be taken as real. For, we know that such a constant serves merely to change the scale of measurement in virtue of its modulus and to rotate the figure in virtue of its amplitude. Since the positive u-axis has been mapped onto the positive v-axis, the figure as a whole has not been rotated but merely fanned out and consequently the amplitude of C may be taken as zero, whence this constant is real. Since the scale of measurement will depend upon the data of our problem and upon the system of units employed, we have C as an arbitrary real constant.

Therefore, separating real and imaginary components in (9.15), we get

$$V = C\,(u^2 - v^2),\ S = 2\,C\,uv. \tag{9.16}$$

It follows that the maps of V = constant are rectangular hyperbolas asymptotic to the line v = u and the maps of s = const. are also rectangular hyperbolas asymptotic to the u- and v-axes. Thus the full curves in Fig.17 represent streamlines and the dotted curves equipotentials in the neighbourhood of a corner of the sheet.

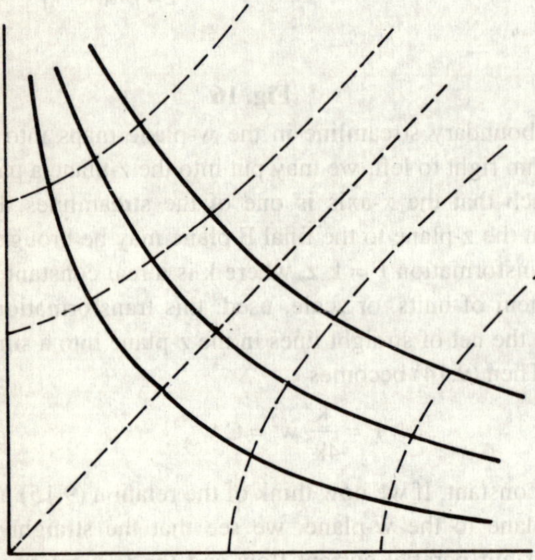

Fig. 17

In the same way, as above, we may interpret our result in connection with the analogous problem of heat flow for which Equation (9.16) give the flux lines and isotherms near a corner, or in connection with fluid flow, for which Equation (9.16) represents flux lines and the lines of velocity potential.

Example 2. Flow Near Two 90° Corners

Now we deal with a rectangular strip of width a and so long compared to a that its length may be supposed infinite. It can be called a semi-infinite channel in fluid Mechanics applications. We again place our sheet in the first quadrant of the w-plane as shown in Fig. 18 and

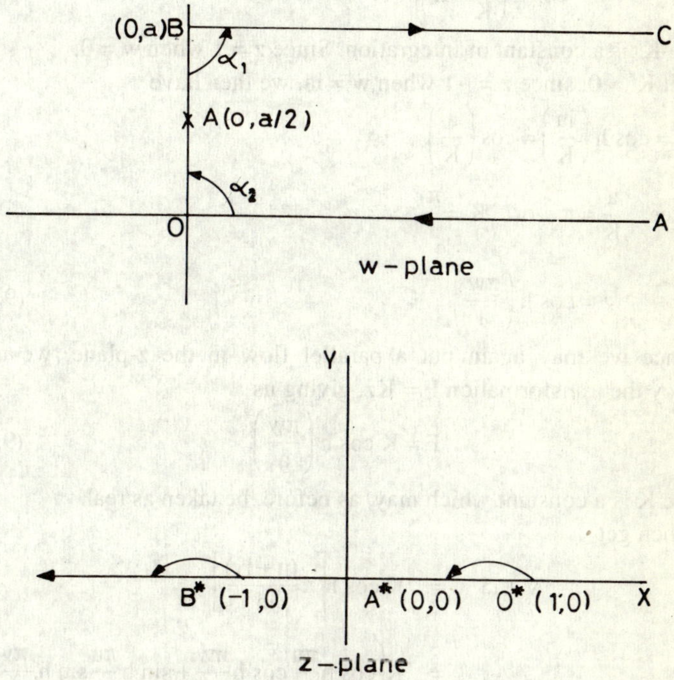

Fig. 18

suppose that one streamline is along UOBC as indicated by the arrows. We map this polygon with interior angles $\alpha_1 = \pi/2$, $\alpha_2 = \pi/2$, symmetrically on to the x-axis, the midpoint A(0, a/2) of the edge OB mapping onto the origin of the z-plane and the corners O and B mapping onto the points x = 1, x = −1, respectively. The stream boundary UOBC is thus mapped on to the x-axis, this later streamline proceeding from right to left.

We then have for the Schwarz-Christoffel transformation

$$\frac{dw}{dz} = K\,(z+1)^{-\frac{1}{2}}\,(z-1)^{-\frac{1}{2}} = \frac{K}{(z^2-1)^{\frac{1}{2}}} \tag{9.17}$$

whence $w = K\,(\cos h^{-1} z + K')$

or $$z = \cos h\left(\frac{w}{K} - K'\right)$$

where K' is a constant of integration. Since $z = 1$ when $w = 0$, we get $K' = 0$; since $z = -1$ when $w = ia$, we then have

$$-1 = \cos h\left(\frac{ia}{K}\right) = \cos\left(\frac{a}{K}\right),$$

whence $\frac{a}{K} = \pi$ or $K = \frac{\pi}{a}$

and $$z = \cos h\left(\frac{\pi w}{a}\right), \tag{9.18}$$

Since we may again put a parallel flow in the z-plane, we may employ the transformation $F = Kz$, giving us

$$F = K \cos h\left(\frac{\pi w}{a}\right), \tag{9.19}$$

where K is a constant which may, as before, be taken as real. We then get

$$\begin{aligned} V + iS &= K \cos h\left[\pi \frac{(u+iv)}{a}\right] \\ &= K\left(\cos h \frac{\pi u}{a} \cos h \frac{i\pi v}{a} + \sin h \frac{\pi u}{a} \sin h \frac{i\pi v}{a}\right) \\ &= K\left(\cos h \frac{\pi u}{a} \cos \frac{\pi v}{a} + i \sin h \frac{\pi u}{a} \sin \frac{\pi v}{a}\right) \end{aligned}$$

and, therefore,

$$\left.\begin{aligned} V &= K \cos h \frac{\pi u}{a} \cos \frac{\pi v}{a}, \\ S &= K \sin h \frac{\pi u}{a} \sin \frac{\pi v}{a}. \end{aligned}\right\} \tag{9.20}$$

The maps of V = constant will be the equipotentials (or isotherms) and the maps of S = constant will be the streamlines.

Example 3. Edge Effect in a Condenser: Flow Out of a Channel.

Let the edges of a parallel-plate condenser, or of a channel conveying a liquid lie along the lines v = ± a to the left of the points (-b, ± a) in the second and third quadrants. We suppose these edges so long compared to the distance 2a between them that they may be considered infinite in length.

If the plates v = ± a of a condenser are at different potentials, there will be lines of force between them. Sufficiently far to the left of u = –b, these lines will be sensibly straight and vertical, but as we approach the boundaries at u = –b they will evidently be considerably curved. We wish to study this edge effect for the parallel-plate condenser.

Another physical interpretation for the present problem is in connection with fluid flow. Again, far to the left of u = –b, the streamlines will be nearly horizontal straight lines, but will curve as we proceed to the right and diverge as the liquid issues from the channel.

Because of the evident symmetry in the u-axis for both physical problems, we need consider only the upper half of the figure in the w-plane (Fig. 19). We shall then think of our open polygon as running from A(u = –∞) to P(–b, a), then back along PA and joining onto B at u = –∞, and then to the right along u-axis. Thus the angles involved are $\alpha_1 = 2\pi$ at P and $\alpha_2 = 0$ at B.

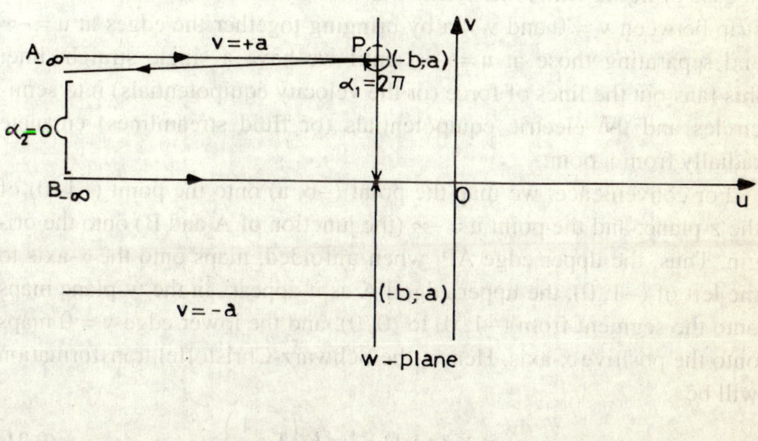

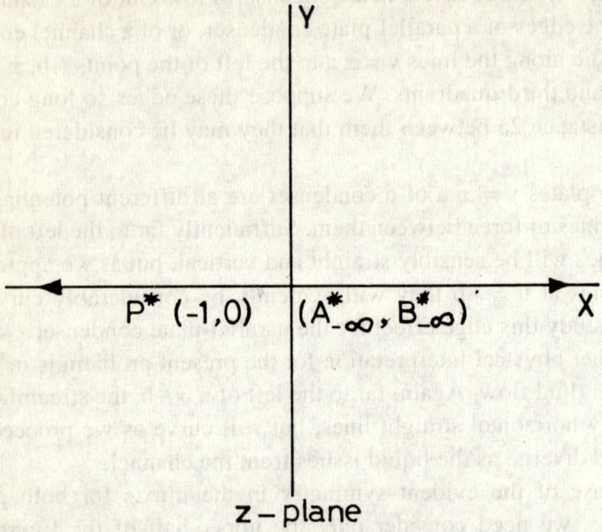

Fig. 19

To open out this polygon, we first unfold the line AP by rotating it through 180° about P; this gives us flow in a uniform strip and thus straightens out the equipotentials of the electric field (or the streamlines in case of liquid flow) into horizontal straight lines. We then fan out the strip between $v = 0$ and $v = a$ by bringing together the edges at $u = -\infty$ and separating those at $u = +\infty$, until we have a single straight line; this fans out the lines of force (or the velocity equipotentials) into semi-circles and the electric equipotentials (or fluid streamlines) emanate radially from a point.

For convenience, we map the point $(-b, a)$ onto the point $(-1, 0)$ of the z-plane, and the point $u = -\infty$ (the junction of A and B) onto the origin. Thus, the upper edge AP, when unfolded, maps onto the x-axis to the left of $(-1, 0)$, the upper edge PA as it appears in the w-plane maps onto the segment from $(-1, 0)$ to $(0, 0)$, and the lower edge $v = 0$ maps onto the positive x-axis. Hence, the Schwarz-Christoffel transformation will be

$$\frac{dw}{dz} = K\,(z+1)\,z^{-1} = K\left(1 + \frac{1}{z}\right), \qquad (9.21)$$

and, therefore,

$$w = K(z + \ln z) + K'. \qquad (9.22)$$

Now the fluid streamlines along the boundary of the w-polygon run from left to right along AP and the u-axis, and their maps run away from the origin along the x-axis. Here, therefore, we may put into the z-plane a radial flow emnating from z = 0. To transform this radial flow into parallel flow, we take the mapping function $z = e^{kF}$. Since the real constant k serves merely to change the scale of our map, giving us kV and kS instead of simply V and S, we set k = 1 for simplicity. Then we have

$$w = K(e^F + F) + K'. \qquad (9.23)$$

To determine the constants K and K′, write z in polar form, $z = r\,e^{i\theta}$, so that Equation (9.22) becomes

$$w = K(re^{i\theta} + \ln r + i\theta) + K'. \qquad (9.24)$$

As we pass from v = 0 to v = a in the w-plane, w changes by ia, and we may think of z as passing from a point on the positive x-axis, where $\theta = 2n\pi$ and n is an integer or zero, to a point on the negative x-axis, where $\theta = (2n + 1)\pi$ and then allow these points to approach the origin. For these changes, we, therefore, have

$$w]_0^{ia} = \lim_{r \to 0} \; [K(r\,e^{i\theta} + \ln r + i\theta)]_{\theta = 2n\pi}^{\theta = (2n+1)\pi}$$

$$ia = \lim_{r \to 0} \; [K(-2r + i\pi)] = i\pi K$$

whence

$$K = a/\pi. \qquad (9.25)$$

Moreover, z = −1 when w = −b + ia, so that since $\ln(-1) = \ln 1 + (2n + 1)i\pi = (2n + 1)i\pi$, we get from (9.24) and (9.25)

$$-b + ia = \frac{a}{\pi} \; [-1 + (2n + 1)i\pi] + K',$$

$$K' = \frac{a}{\pi} - b - 2nia.$$

Consequently our transformation takes the form

$$w = \frac{a}{\pi}(e^F + F) + \frac{a}{\pi} - b - 2nia$$

so that

$$u = \frac{a}{\pi}(e^V \cos S + V + 1) - b,$$

$$v = \frac{a}{\pi}(e^V \sin S + S) - 2na.$$

Now the line v = 0 maps onto the line y = 0 and since, from $z = e^F$, we have $y = e^V \sin S$, it follows that v = 0 corresponds to S = 0. Hence, in the above expression for v, we must take n = 0, and, therefore, we get as our final result,

$$u = \frac{a}{\pi}(e^V \cos S + V + 1) - b,$$

$$v = \frac{a}{\pi}(e^V \sin S + S). \tag{9.26}$$

Letting v = constant in (9.26), we obtain equations for u and v in terms of the parameter S, from which we may plot various lines of force of the electrostatic field near the edges of the condenser. By putting S = const. in (9.26), we get equations for u and v in terms of the parameter V, which enable us to plot the lines of flow of liquid from the channel.

Example 4. Point Source and Point Sink.

As our next problem, we investigate the distribution of streamlines and equipotentials in a sheet of conducting material into which current is fed at a point A and out of which it flows at a point B. The points A and B are called, respectively, source and sink.

Here, we put the known pattern of streamlines and equipotentials for parallel current flow in the w-plane and transform this into the desired pattern in the z-plane (Fig. 20).

Let the source and sink be, respectively, located at A(a, 0) and B(−a, 0) of the z-plane, and let A_1 B_1 and A_2 B_2 be boundaries of the strip in the w-plane along which parallel flow takes place. Split the upper edge A_2B_2 at its centre C − C′, rotate the right-hand half about A_2 in the clockwise and the left-hand half about B_2 in the counterclockwise direction, and bring A_2 toward A_1 and B_2 toward B_1. Let (a, 0) in the z-plane correspond to the junction of A_1 and A_2, and let (−a, 0)

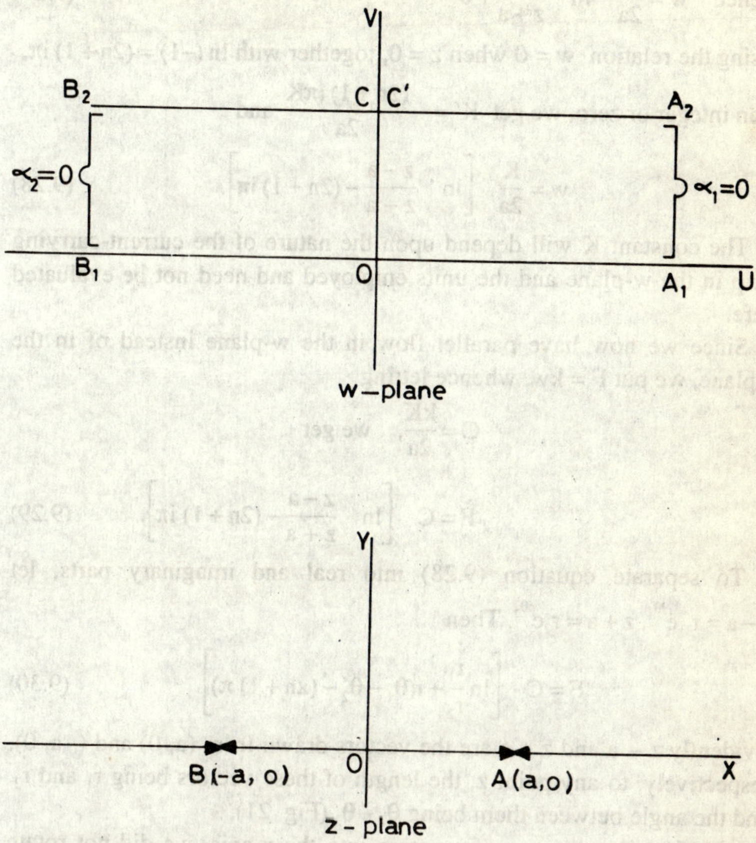

Fig. 20

correspond to the junction of B_1 and B_2. Map the origin in the w-plane onto the origin in the z-plane and let C go into $x = -\infty$ and C′ into $x = +\infty$.

The internal angles, α_1 at $A_1 - A_2$ and α_2 at $B_1 - B_2$, being both zero, the Schwarz-Christoffel transformation will be

$$\frac{dw}{dz} = K\,(z-a)^{-1}\,(z+a)^{-1} = \frac{K}{z^2 - a^2}.$$

Hence $w = \frac{K}{2a} \ln \frac{z-a}{z+a} + \theta$ (9.27)

Using the relation $w = 0$ when $z = 0$, together with $\ln(-1) = (2n+1)i\pi$, n an integer or zero, we get $K' = -\frac{(2n+1)i\pi K}{2a}$ and

$$w = \frac{K}{2a}\left[\ln \frac{z-a}{z+a} - (2n+1)i\pi\right]. \quad (9.28)$$

The constant K will depend upon the nature of the current-carrying strip in the w-plane and the units employed and need not be evaluated here.

Since we now have parallel flow in the w-plane instead of in the z-plane, we put $F = kw$, whence letting

$$C = \frac{kK}{2a}, \text{ we get}$$

$$F = C\left[\ln \frac{z-a}{z+a} - (2n+1)i\pi\right]. \quad (9.29)$$

To separate equation (9.28) into real and imaginary parts, let $z - a = r_1 e^{i\theta_1}$, $z + a = r_2 e^{i\theta_2}$. Then

$$F = C\left[\ln \frac{r_1}{r_2} + i(\theta_1 - \theta_2 - (2n+1)\pi)\right]. \quad (9.30)$$

Evidently $z - a$ and $z + a$ are the vectors drawn from (a, 0) and (–a, 0), respectively, to any point z, the length of these vectors being r_1 and r_2, and the angle between them being $\theta_1 - \theta_2$ (Fig. 21).

In transforming the w-configuration into the x-axis, we did not rotate the figure, and, therefore, C may be taken as real. Consequently, we get

$$V = C \ln \frac{r_1}{r_2}, \; S = C\,[\theta_1 - \theta_2 - (2n+1)\pi]. \quad (9.31)$$

Now any point of segment $A_1 B_1$ in the w-plane, for which $v = 0$ maps onto a corresponding point of the segment AB in the z-plane, for which $\theta_1 = \pi$, $\theta_2 = 0$. Hence, the second of Equation (9.31) gives us $n = 0$, whence we have

$$V = C \ln \frac{r_1}{r_2}, \; S = C\,(\theta_1 - \theta_2 - \pi). \quad (9.32)$$

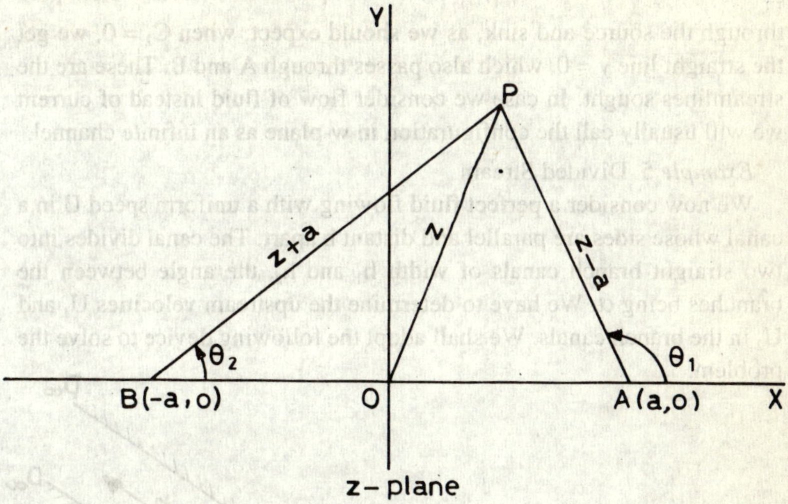

Fig. 21

From the first of Equation (9.31), we see that V will be constant if and only if the ratio r_1/r_2 is constant. Thus the maps of the equipotentials V = constant in the w-plane will be given by

$$\frac{r_1}{r_2} = \frac{\sqrt{(x-a)^2+y^2}}{\sqrt{(x+a)^2+y^2}} = C_1$$

or $$(1-c_1^2)(x^2+y^2) - 2a(1+c_1^2)x + a^2(1-c_1^2) = 0, \tag{9.33}$$

where C_1 is a constant, which must be taken as positive since r_1/r_2 is positive. For $C_1 \neq 1$, (9.33) represents circles with their centres on the x-axis: when $C_1 = 1$, we get, as could be anticipated from $r_1/r_2 = 1$, the straight line x = 0. These, then, are equipotentials of our problem.

Corresponding to the streamlines S = constant, we get

$$\tan\left(\frac{S}{c} + \pi\right) = \tan(\theta_1 - \theta_2)$$

$$= \frac{\dfrac{y}{x-a} - \dfrac{y}{x+a}}{1 + \dfrac{y^2}{x^2-a^2}} = \frac{2ay}{x^2+y^2-a^2} = C_2,$$

a constant, or $$C_2(x^2+y^2) - 2ay - C_2a^2 = 0. \tag{9.34}$$

When $C_2 \neq 0$, Equation (9.34) also represents circles, which pass through the source and sink, as we should expect; when $C_2 = 0$, we get the straight line $y = 0$, which also passes through A and B. These are the streamlines sought. In case we consider flow of fluid instead of current we will usually call the configuration in w-plane as an infinite channel.

Example 5. Divided Stream

We now consider a perfect fluid flowing with a uniform speed U in a canal whose sides are parallel and distant h apart. The canal divides into two straight branch canals of width h_1 and h_2, the angle between the branches being α. We have to determine the upstream velocities U_1 and U_2 in the branch canals. We shall adopt the following device to solve the problem.

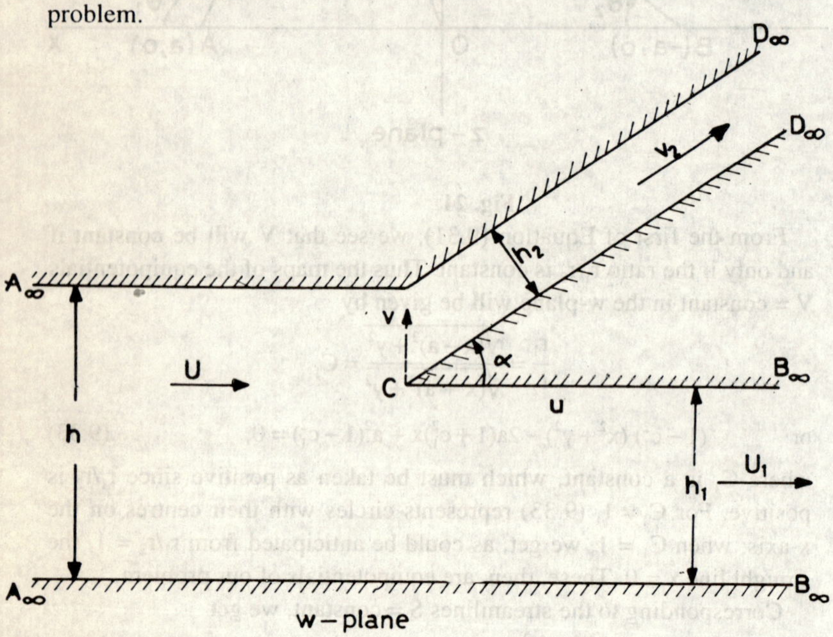

Fig. 22

At A_∞ the fluid can be considered to be created from a source at a rate Uh. Similarly, at B_∞ and D_∞ there exist sinks where the fluid is destroyed at rates U_1h_1, U_2h_2, respectively. Under a conformal transformation, sources and sinks transform into sources and sinks, respectively. From the above arguments we obviously have

$$Uh = U_1h_1 + U_2h_2 \tag{9.35}$$

which is followed by the equation of continuity.

We apply the Schwarz-Christoffel transformation to transform the interior of the domain $A_\infty B_\infty CD_\infty EA_\infty$ into the upper half or z-plane in such a manner that the point E in the w-plane correspond to ∞ in the z-plane: C, the origin, in w-plane correspond to $z = 0$ in the z-plane and A_∞ in w-plane is mapped upon the point $z = -a$ in the z plane (Fig. 23).

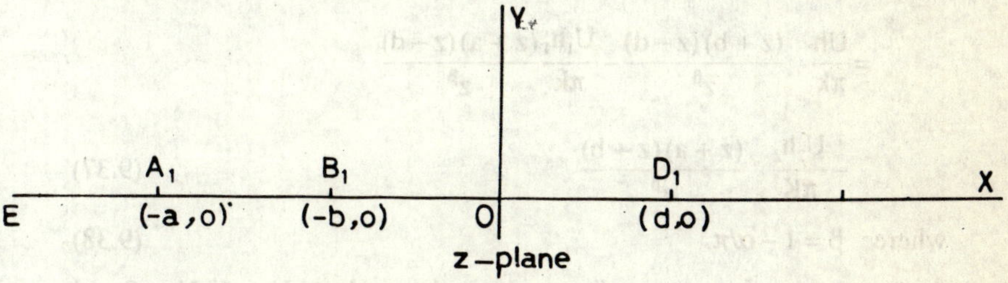

Fig. 23

We assume here as an approximation to real flow that the point C is a stagnation point, *i.e.*, a point at which the velocity of stream is zero. We further suppose that B_∞, D_∞ in the w-plane are mapped onto the points B_1 $(-b, 0)$ and D_1 $(d, 0)$ in z-plane, respectively, where b, d can be determined in terms of a. We note here that the outer boundary corresponds to negative values of z only.

The Schwarz-Christoffel transformation then gives

$$\frac{dw}{dz} = K(z + a)^{-1}(z + b)^{-1}(z - 0)^{1-\frac{\alpha}{\pi}}(z - d)^{-1} \tag{9.36}$$

since the interior angles of the polygonal boundary $A_\infty B_\infty CD_\infty EA_\infty$ are $0, 0, 2\pi - \alpha, 0$, there being no contribution from the vertex E, since E corresponds to infinity.

From the previous remarks there exists a source at A_1 in the z-plane at which fluid is created at a rate Uh over an angle π, and so is equivalent to a source of strength $\frac{Uh}{\pi}$. Similarly at B_1 and D_1 in the z-plane there are point sinks of strengths $\frac{U_1h_1}{\pi}$ and $\frac{U_2h_2}{\pi}$. We can, thus, write the complex

$$\Omega = \frac{Uh}{\pi}\log(z+a) - \frac{U_1h_1}{\pi}\log(z+b) - \frac{U_2h_2}{\pi}\log(z-d).$$

so

$$\frac{d\Omega}{dz} = V_x - iV_y = \frac{Uh}{\pi(z+a)} - \frac{U_1h_1}{\pi(z+b)} - \frac{U_2h_2}{\pi(z-d)}$$

whence from (9.36)

$$\frac{d\Omega}{dw} = V_u - iV_v = \frac{d\Omega}{dz}\cdot\frac{dz}{dw}$$

$$= \frac{Uh}{\pi k}\,\frac{(z+b)(z-d)}{z^{\beta}} - \frac{U_1h_1}{\pi k}\,\frac{(z+a)(z-d)}{z^{\beta}}$$

$$- \frac{U_2h_2}{\pi K}\,\frac{(z+a)(z+b)}{z^{\beta}} \qquad (9.37)$$

where $\beta = 1 - \alpha/\pi.$ (9.38)

At $z = -a = ae^{i\pi}$ corresponding to A_∞ in the w-plane, $V_u = U, V_v = 0$ and so

$$U = \frac{Uh}{\pi K}\,\frac{(a-b)(a+d)}{a^{\beta}e^{i\pi\beta}} \qquad (9.39)$$

At $z = -b = b\,e^{i\pi}$ corresponding to E_∞ in the w-plane,

$V_u = U_1, V_v = 0$, and so

$$U_1 = \frac{U_1h_1}{\pi K}\,\frac{(a-b)(b+d)}{b^{\beta}e^{i\pi\beta}} \qquad (9.40)$$

At $z = d = de^{i0}$ corresponding to D_∞ in the w-plane,

$V_u = U_2\cos\alpha, \quad V_v = U_2\sin\alpha$, and so

$$V_u - iV_v = U_2e^{-i\alpha} = -\frac{U_2h_2}{\pi K}\,\frac{(a+d)(b+d)}{d^{\beta}}.$$

But $-i\alpha = i\pi\beta - \pi i$, and since $e^{-\pi i} = -1$,

$$U_2e^{i\pi\beta} = \frac{U_2h_2}{\pi K}\,\frac{(a+d)(b+d)}{d^{\beta}}. \qquad (9.41)$$

Equations (9.39), (9.40) and (9.41) can now be written as

$$\pi k e^{i\pi\beta} = \frac{h(a-b)(a+d)}{a^\beta} = \frac{h_1(a-b)(b+d)}{b^\beta}$$

$$= h_2 \frac{(a+d)(b+d)}{d^\beta}$$

which implies that $K\,e^{i\pi\beta}$ is a real constant and

$$\frac{a+d}{b+d} = \frac{h_1}{h}\left(\frac{a}{b}\right)^\beta, \frac{a-b}{b+d} = \frac{h_2}{h}\left(\frac{a}{d}\right)^\beta \tag{9.42}$$

At origin C in the w-plane there is a stagnation point, and so $\frac{d\Omega}{dw}$ is zero. Hence, since w = 0 corresponds to z = 0, it follows from (9.37) that the coefficient of $z^{-\beta}$ is zero, *i.e.*,

$$-Uhbd + U_1h_1\,ad - U_2h_2ab = 0$$

or

$$\frac{Uh}{a} = \frac{U_1h_1}{b} - \frac{U_2h_2}{d} \tag{9.43}$$

From (9.35) and (9.43), we have

$$\frac{U_1}{U} = \frac{h}{h_1}\frac{b(a+d)}{a(b+d)}, \frac{U_2}{U} = \frac{h}{h_2}\frac{d(a-b)}{a(b+d)}$$

or from (9.42)

$$\frac{U_1}{U} = \left(\frac{b}{a}\right)^{1-\beta} = \left(\frac{b}{a}\right)^{\alpha/\pi}, \frac{U_2}{U} = \left(\frac{d}{a}\right)^{1-\beta} = \left(\frac{d}{a}\right)^{\alpha/\pi}$$

$$i.e., \quad \frac{b}{a} = \left(\frac{U_1}{U}\right)^{\pi/\alpha}, \frac{d}{a} = \left(\frac{U_2}{U}\right)^{\pi/\alpha} \tag{9.44}$$

Writing (9.35) in the form

$$1 = \frac{U_1h_1}{Uh} + \frac{U_2h_2}{Uh} \tag{9.45}$$

and introducing

$$\frac{h_1}{h} = \lambda, \frac{h_2}{h} = \mu, \frac{U_1h_1}{Uh} = \upsilon \tag{9.46}$$

we have from (9.45)

$$\frac{U_2h_2}{Uh} = 1 - \upsilon. \tag{9.47}$$

Finally (9.43) is equivalent to

$$1 = \frac{U_1h_1}{Uh}\left(\frac{a}{b}\right) - \frac{U_2h_2}{Uh}\left(\frac{a}{d}\right)$$

which with (9.44), (9.46) and (9.47) gives

$$1 = \lambda\left(\frac{\upsilon}{\lambda}\right)^{1-\pi/\alpha} - \mu\left(\frac{1-\upsilon}{\mu}\right)^{1-\pi/\alpha} \tag{9.48}$$

a transcendental equation for υ and hence for U_1 and U_2.

Example 6. Uniform Flow Past A Plate

We consider an infinite plate of width 2a, placed at right angles to a uniform stream with speed * U_0 (Fig. 24) in the w-plane.

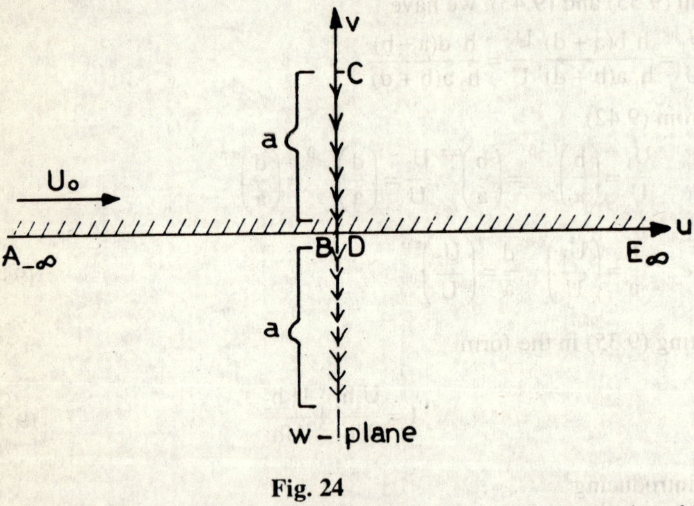

Fig. 24

By symmetry the u-axis is a streamline and so it can be replaced by a rigid boundary. Thus we can consider first the streaming motion over a

*The problem is realistic for low speed of stream only; for otherwise there will then exist a constant pressure wake and a region of dead water behind the plate.

horizontal bed having a thin wall of height at right angles to the direction of stream, then by reflection in the u-axis the proposed problem can be solved.

To transform the polygonal line $A_{-\infty}BCDE_{\infty}$ in the w-plane into the real axis in z-plane we consider $A_{-\infty}, E_{\infty}$ to correspond to $\bar{A}_{-\infty}, \bar{E}_{\infty}$ in the z-plane: B and D to correspond to $\bar{B}$ (–a, 0) and $\bar{D}$ (a, 0), respectively, in z-plane and the point C (0, ai) to transform into $\bar{C}$, the origin, in z-plane (Fig. 25).

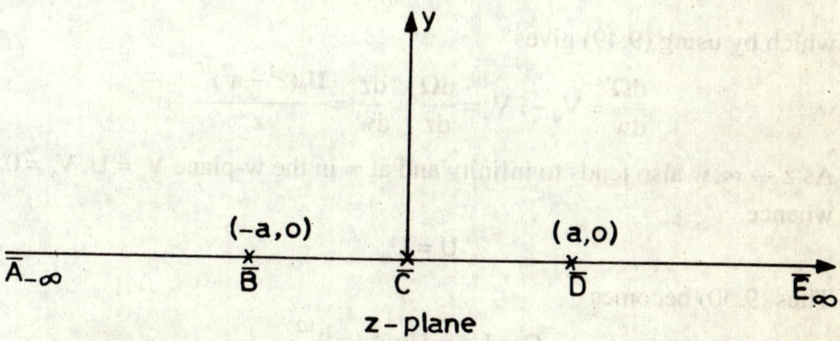

Fig. 25

The angles at B, C, D are $\pi/2$, 2π, $\pi/2$, respectively. Thus, the Schwarz-Christoffel transformation takes the form

$$\frac{dw}{dz} = K(z+a)^{-1/2}\,(z-0)^{1}\,(z-a)^{-1/2}$$

$$= \frac{KZ}{(z^2-a^2)^{1/2}},$$

which on integration gives

$$w = K(z^2-a^2)^{1/2} + K',$$

K′ being the constant of integration which is zero since w = 0 when z = ± a. Further since w = ai when z = 0, the above equation gives.

$$ai = K\,ai$$

i.e. $$K = 1$$

and the transformation can be written as

$$\left.\begin{aligned} w &= (z^2 - a^2)^{1/2} \\ \text{or} \qquad z &= (w^2 + a^2)^{1/2} \end{aligned}\right\} \tag{9.49}$$

The flow in w-plane can be considered as due to a source at $A_{-\infty}$ and a sink of equal strength at E_∞. Under the transformation, since $A_{-\infty}, E_\infty$ transform into the source, and sink at $\bar{A}_{-\infty}, \bar{E}_\infty$, respectively, the flow in z-plane is also uniform and is parallel to x-axis with speed U_0 (say). Thus, the complex potential in z-plane is given by

$$\Omega = U_0 z, \tag{9.50}$$

which by using (9.49) gives

$$\frac{d\Omega}{dw} = V_u - i\,V_v = \frac{d\Omega}{dz}\;\frac{dz}{dw} = \frac{U_0(z^2 - a^2)^{1/2}}{z}$$

As $z \to \infty$, w also tends to infinity and at ∞ in the w-plane $V_u = U$, $V_v = 0$ whence,

$$U = U_0.$$

Thus (9.50) becomes

$$\Omega = Uz = U(w^2 + a^2)^{1/2},$$

which gives the complex potential in the w-plane.

Example 7. A Steady State Temperature Distribution Problem

Solution: We shall determine the steady state temperature distribution for $v \geq 0$ if the u-axis is kept at zero temperature and the segment $0 < v \leq a$ of the v-axis is maintained at a temperature $T = T_0$ (Fig. 26)

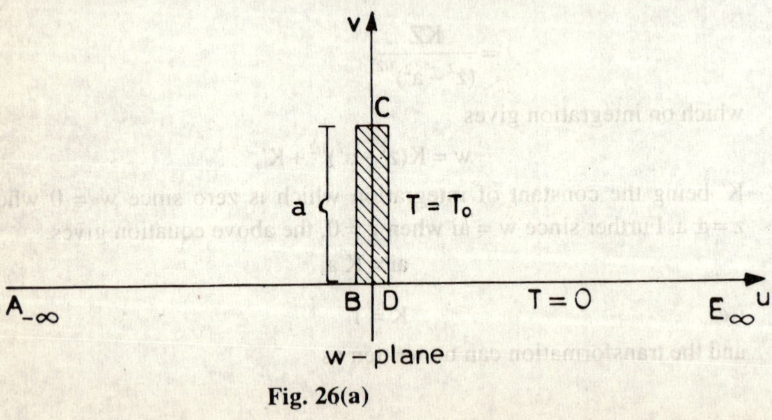

Fig. 26(a)

The transformation of previous example is used to transfoım the u-axis and the segment $0 < v \le a$ into the real-axis $y = 0$ in z-plane.

The steady state temperature satisfies the Laplace's equation

$$\frac{\partial^2 T}{\partial u^2} + \frac{\partial^2 T}{\partial v^2} = 0$$

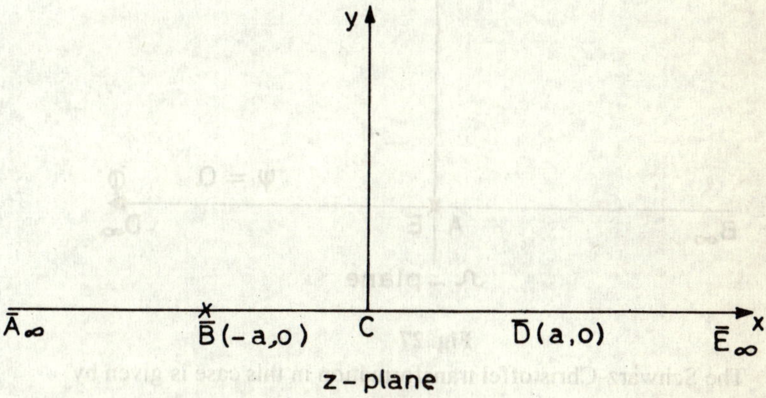

Fig. 26(b)

subject to the boundary conditions

$$\left.\begin{aligned} T &= 0 \text{ for } u < 0 \\ T &= T_0 \text{ for } u = 0,\ 0 < v \le a \\ T &= 0 \quad \text{for} \quad u > 0, \end{aligned}\right\} \tag{9.51}$$

there being discontinuities in temperature at B and D.

If, $\Omega = \phi + i\psi$ is the complex potential in the w-plane with ψ = con stant representing isothermals, then since Ω is a regular function of z, $\psi = T$ automatically satisfies Laplace's equation. The boundary conditions for ψ are same as for T. From this, we shall determine the Ω.

In (ϕ, ψ)-plane (Fig. 27) $\psi = 0$ whatever may be the value of ϕ on isothermal $BA_{-\infty}E_{\infty}D$, B and D being discontinuous points. Along the isothermal BCD, $\psi = T_0$ and hence Fig. 26 represents the Ω-plane. We shall now transform this plane into the z-plane making B_{∞}, D_{∞} correspond to $z = -a$, $z = a$, respectively.

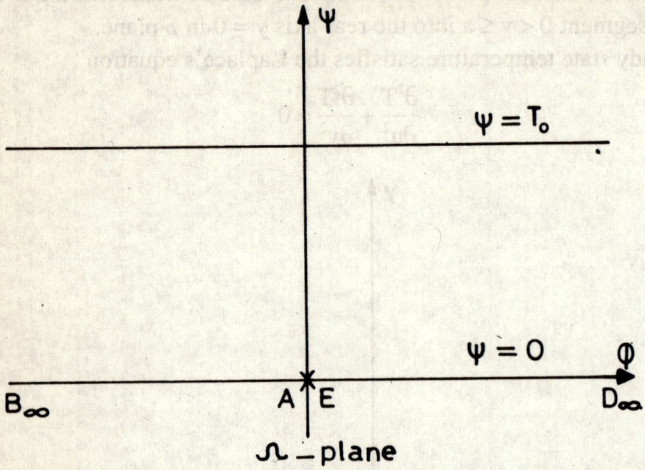

Fig. 27

The Schwarz-Christoffel transformation in this case is given by

$$\frac{d\Omega}{dz} = K(z+a)^{-1}(z-a)^{-1}$$

since the angles at B_∞ D_∞ are zero. Integrating this equation, we get

$$\Omega = \frac{K}{2a} \quad \{\log(z-a) - \log(z+a)\} + K'$$

where K′ is the constant of integration. Since A_∞, E_∞ in z-plane lie on the curve $\phi = 0$ and from Fig 27, $\psi = 0$ for those points, Ω is also zero. Thus in the above equation, $\Omega \to 0$ as $z \to \infty$ and $K' = 0$.

Thus

$$\Omega = \frac{K}{2a} \log \frac{z-a}{z+a}$$

$$= \frac{K}{2a} \left\{ \log \left| \frac{z-a}{z+a} \right| + i \arg \frac{z-a}{z+a} \right\}. \quad (9.52)$$

Now we shall evaluate the constant K. If z lies on real-axis then

for $x < -a$, $\arg(z-a) = \pi$, $\arg(z+a) = \pi$, $\arg \dfrac{z-a}{z+a} = 0$

for $-a < x < a$, $\arg(z-a) = \pi$, $\arg(z+a) = 0$, $\arg \dfrac{z-a}{z+a} = \pi$

for $x > a$, $\arg(z-a) = 0$, $\arg(z+a) = 0$, $\arg \dfrac{z-a}{z+a} = 0$,

and so,

for $\quad x < -a, \quad = \quad + i.0 = \dfrac{K}{2a} \log \dfrac{z-a}{z+a}$

which implies that K is real.

for $\quad -a < x < a$,

$$\text{Im}(\Omega) = \psi = T_0 = \frac{K\pi}{2a}$$

whence

$$\frac{K}{2a} = \frac{T_0}{\pi}.$$

Thus the complex potential is

$$\Omega = \frac{T_0}{\pi} \log \frac{z-a}{z+a}$$

with

$$T = \psi = \frac{T_0}{\pi} \quad \{\arg(z-a) - \arg(z+a)\} = \text{constant.} \tag{9.53}$$

giving the isothermals.

We now return to the w-plane by means of the transformation (9.49) of previous example. To this end, let

$\arg(z-a) = \gamma_1$, $\arg(z+a) = \gamma_2$, then

$$\tan \gamma_1 = \frac{y}{x-a}, \tan \gamma_2 = \frac{y}{x+a}$$

$$\tan(\gamma_1 - \gamma_2) = \frac{2ay}{x^2+y^2-a^2}.$$

Thus from (9.53), we have

$$T = \frac{T_0}{\pi} \tan^{-1} \frac{2ay}{x^2+y^2-a^2}. \tag{9.54}$$

Equating real and imaginary parts of (9.49), we get

$$u^2 - v^2 = x^2 - y^2 - a^2,$$

$$uv = xy.$$

From these equations, taking the only admissible values, we obtain

$$x^2 = \frac{1}{2}\left\{(u^2 - v^2 + a^2) + \sqrt{(u^2 - v^2 + a^2)^2 + 4u^2v^2}\right\},$$

$$y^2 = \frac{1}{2}\left\{\sqrt{(u^2 - v^2 + a^2)^2 + 4u^2v^2} - (u^2 - v^2 + a^2)\right\}.$$

Substitution of these values in (9.54) gives the required solution in w-plane as

$$T = \frac{T_0}{\pi}\tan^{-1}\left[\frac{\left\{2\left[\sqrt{(u^2 - v^2 + a^2)^2 + 4u^2v^2} - (u^2 - v^2 + a^2)\right]\right\}}{\sqrt{(u^2 - v^2 + a^2)^2 + 4u^2v^2} - a^2}\right]^{1/2} \quad (9.55)$$

The solution of the electrical problem where the wall BC is at a potential V_0 whilst the axis $v = 0$ is at potential zero, is exactly the same as that of the heat problem given.

The similar problem in hydrodynamics can be easily deduced. The complex potential due to a source and sink of equal strengths at $z = a$ and $z = -a$, respectively, is

$$\Omega = \frac{T_0}{\pi}\log\frac{z - a}{z + a}. \quad (9.56)$$

The source and sink in z-plane transform under $w = (z^2 - a^2)^{1/2}$ into

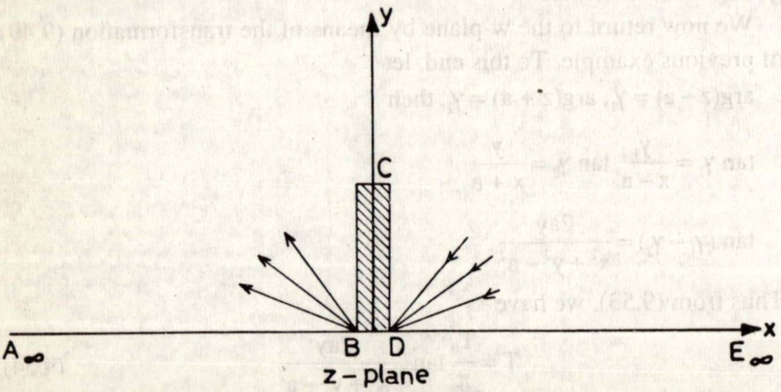

Fig. 28

a source and sink at B and D, respectively. Thus the complex potential in the w-plane is that for the following configuration:

Example 8. Electrode And Plate

We consider a semi-infinite electrode of depth 2b kept at a potential V_0 and a semi-infinite plate at zero potential placed at a distance c from the electrode (Fig. 29). We have to determine the complex potential in this case.

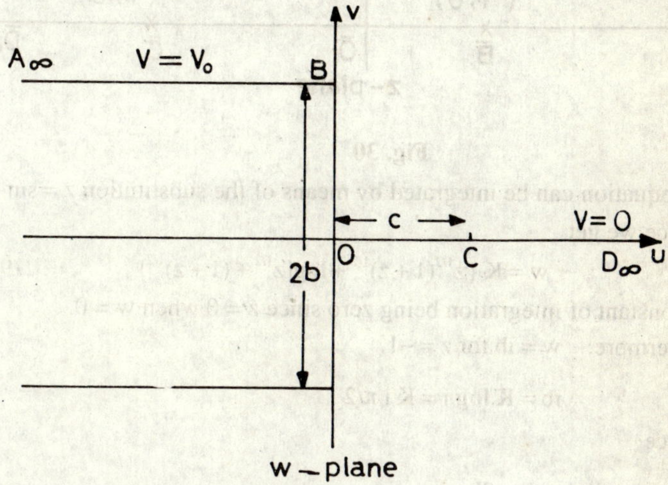

Fig. 29

We consider first the problem in the upper-half of w-plane and then by reflection in u-axis we obtain the required solution. We transform the polygon $A_\infty BOCD_\infty$ in w-plane into the real-axis in z-plane in such a manner that A_∞, D_∞ transform to the points $\overline{A}_\infty$, $\overline{D}_\infty$ in z-plane; w = 0, w = ib, w = c transform into z = 0, z = −1, z = γ, in z-plane (Fig. 30), respectively, where γ is some constant. Since the angles at B and O are 3π/2, π/2, respectively, the Schwarz-Christoffel transformation is then given by

$$\frac{dw}{dz} = K(z+1)^{1/2}(z-0)^{-1/2}$$

$$= K\left(\frac{z+1}{z}\right)^{1/2} \qquad (9.57)$$

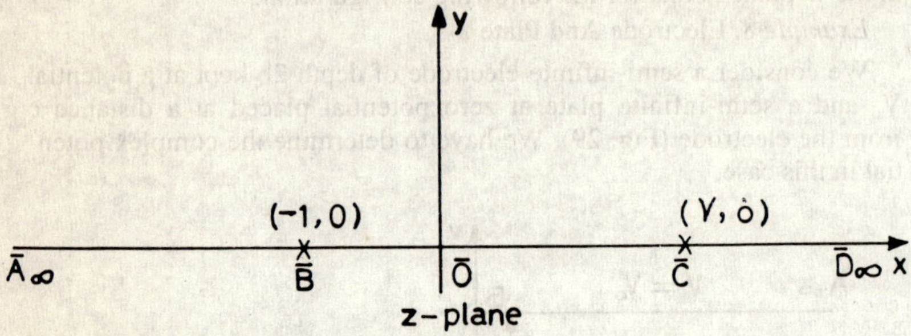

Fig. 30

This equation can be integrated by means of the substitution $z = \sinh^2 \chi$ whence we get

$$w = K\{z^{1/2}(1+z)^{1/2} + \log[z^{1/2} + (1+z)^{1/2}]\} \tag{9.58}$$

the constant of integration being zero since $z = 0$ when $w = 0$.

Furthermore, $w = ib$ for $z = -1$.

$$ib = K \log i = K\, i\, \pi/2$$

whence

$$K = \frac{2b}{\pi},$$

and (9.58) takes the form

$$w = \left(\frac{2b}{\pi}\right)\{z^{1/2}(1+z)^{1/2} + \log[z^{1/2} + (1+z)^{1/2}]\} \tag{9.59}$$

For $w = c$, we have $z = \gamma$ hence

$$\pi c = 2b\{\gamma^{1/2}(1+\gamma)^{1/2} + \log\ [\gamma^{1/2} + (1+\gamma)^{1/2}]\}$$

which determines the constant γ.

We now consider the complex potential $\Omega = \phi + i\psi$ in w-plane, ϕ being the electrostatic potential. Then along $A_\infty BO, \Omega = V_0 + i\psi$, and along $CD_\infty \Omega = i\psi$ implying that ψ is imaginary. The values of ψ at O and C are equal since OC is a field line. The points A_∞, D_∞ can be taken to lie on the same field line $\psi = \psi_1$ say. Thus in Ω-plane, we have the following figure.

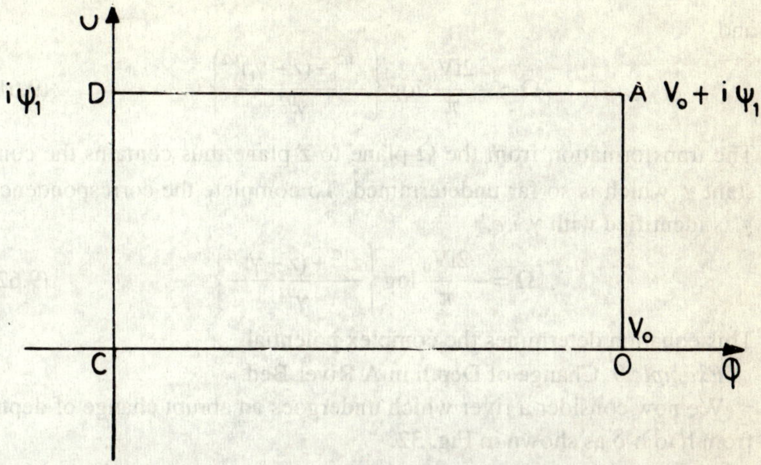

Fig. 31

This rectangle is now transformed into the real-axis of the z-plane in such a manner that the points O, D, A correspond respectively to $\overline{O}, \overline{D}_{\infty}, \overline{A}_{\infty}$. Then the transformation is given by

$$\frac{d\Omega}{dz} = \overline{K}(z-0)^{-1/2}(z-\gamma_1)^{-1/2}$$

$$= \frac{\overline{K}}{\{z(z-\gamma_1)\}^{1/2}}, \tag{9.60}$$

γ_1 being the point in z-plane corresponding to $C(\Omega = 0)$. Making the substitution $w = \gamma_1 \cos h^2\chi$ in (9.60) and then integrating, we get

$$\Omega = 2\overline{K} \cos h^{-1}(z/\gamma_1) = 2\overline{K} \log\left[\frac{z^{1/2} + (z-\gamma_1)^{1/2}}{\gamma_1^{1/2}}\right]$$

the constant of integration being zero since $\Omega = 0$ when $z = \gamma_1$. To evaluate $\overline{K}$ we use the condition

$$\Omega = V_0 \text{ when } z = 0, \quad \text{so that}$$

$$V_0 = \overline{K} \log i = \overline{K}_{\pi i}$$

whence

$$\overline{K} = \frac{V_0}{\pi i}$$

and

$$\Omega = -\frac{2iV_0}{\pi}\log\left\{\frac{z^{1/2}+(z-\gamma_1)^{1/2}}{\gamma_1^{1/2}}\right\} \tag{9.61}$$

The transformation from the Ω-plane to z-plane thus contains the constant γ_1 which is so far undetermined. To complete the correspondence γ_1 is identified with γ, *i.e*,

$$\Omega = -\frac{2iV_0}{\pi}\log\left\{\frac{z^{1/2}+(z-\gamma)^{1/2}}{\gamma^{1/2}}\right\} \tag{9.62}$$

This equation determines the complex potential.

Example 9. Change of Depth in A River Bed

We now consider a river which undergoes an abrupt change of depth from h to h-δ as shown in Fig. 32.

The bottom and upper surfaces of river are the streamlines $\psi = 0$ and $\psi = 1$, respectively. We shall open out the polygonal boundary $A_\infty B_\infty C_\infty DEF_\infty$ into the real-axis in z-plane by making the points B_∞ and C_∞ as coincident and corresponding them with z = 0. The points D and E correspond to (1, 0) and (a, 0), respectively, a being some real number to be determined later.

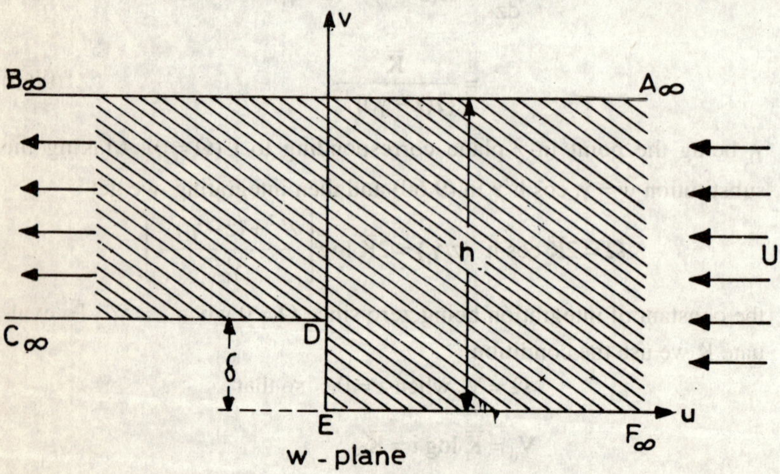

Fig. 32

The Schwarz-Christoffel transformation then gives

$$\frac{dw}{dz} = K\,z^{-1}(z-1)^{-1}(z-a)^{-1} \qquad (9.63)$$

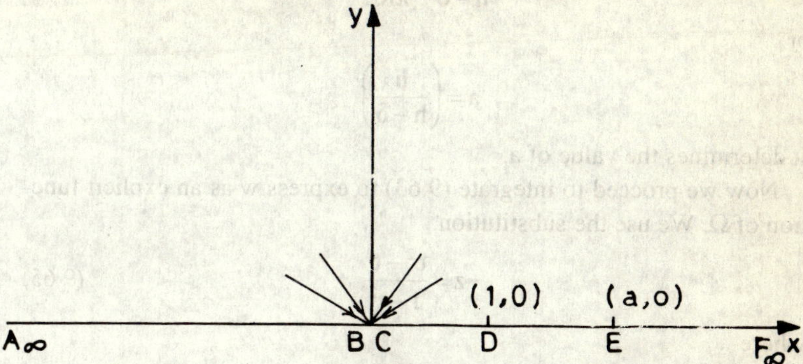

z - plane

Fig. 33

If U is the velocity at A_∞, then velocity at B_∞, by equation of continuity, will be $\frac{Uh}{h-\delta}$. Now in w-plane the flow is from a source of output Uh at A_∞ to a sink of intake Uh at B_∞. Hence in z-plane we have a sink at the origin which in takes the volume Uh per unit time over an angle π. Hence, the strength of sink is

$$\Omega = \frac{Uh}{\pi}\log z \qquad (9.64)$$

whence

$$\frac{d\Omega}{dw} = \frac{d\Omega}{dz}\,\frac{dz}{dw}$$

$$= \frac{Uh}{\pi K}\sqrt{\frac{z-a}{z-a}}$$

Now at $A_\infty(z=\infty)$, $\frac{d\Omega}{dw} = U$. therefore

$$U = \frac{Uh}{\pi K} \quad \text{or } K = h/\pi.$$

Again at $B_\infty(z=0)$, $\frac{d\Omega}{dw} = \frac{Uh}{h-\delta}$. Thererore,

$$\frac{Uh}{h-\delta}=\frac{Uh}{\pi K}\sqrt{a}$$

or

$$a=\left(\frac{h}{h-\delta}\right)^2$$

It determines the value of a.

Now we proceed to integrate (9.63) to express w as an explicit function of Ω. We use the substitution.

$$z=\frac{b^2-t^2}{1-t^2}, \tag{9.65}$$

where

$$b^2=a=\left(\frac{h}{h-\delta}\right)^2$$

and obtain

$$\frac{z-1}{z-b^2}=\frac{1}{t^2}, \frac{dz}{z}=\left(\frac{2t}{1-t^2}-\frac{2t}{b^2-t^2}\right)dt.$$

thus

$$\frac{dw}{dt}=K\left(\frac{2}{1-t^2}-\frac{2}{b^2-t^2}\right)$$

Hence

$$w=\frac{h}{\pi}\left[\log\frac{1+t}{1-t}-\frac{1}{b}\log\frac{b+t}{b-t}\right]+\overline{K} \tag{9.66}$$

where $\overline{K}$ is the constant of integration and is zero since w = 0 when z = a.

Substituting the value of z from (9.65) in (9.64), we obtain

$$\Omega=\frac{Uh}{\pi}\log\frac{b^2-t^2}{1-t^2}$$

Hence

$$t^2=\frac{e^{\Omega\pi/(Uh)}-b^2}{e^{\Omega\pi/(Uh)}-1} \tag{9.67}$$

Thus eliminating t between (9.66) and (9.67) we obtain the equation for complex potential in w-plane.

Applications in Civil-Engineering

In civil engineering applications, the Schwarz-Christoffel transformation is used to study the problems of complicated foundation designs, seepage through earthen dams, etc. In such cases, we transform the complicated foundation designs into a flat floor design and study its flow characteristics. Then by inverting the transformation, we obtain the solution of the desired problem. Below we have discussed several examples to illustrate the method of solving such problems. We shall first make a cursory mention of the theory of seepage flow for general reference.

If a structure is built upon a pervious (permeable) soil and the water level upstream of the structure is higher than it is downstream, enforced filtration or percolation or seepage will occur in the underlying permeable soil. In an earthen dam the seepage will be through the structure as well as the pervious foundation (Fig. 34), while in a weir it will take place only through the foundation (Fig. 35).

The study of the phenomena of seepage flow in structures is important mainly because the design of the apron has to be such as may guard the structure against possible damaging effects of seepage.

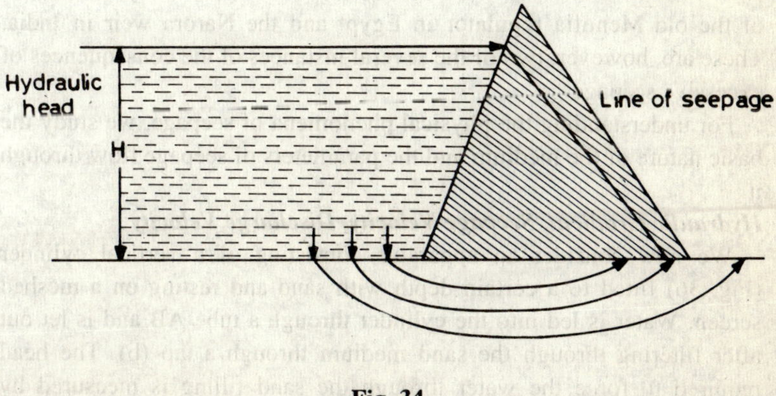

Fig. 34

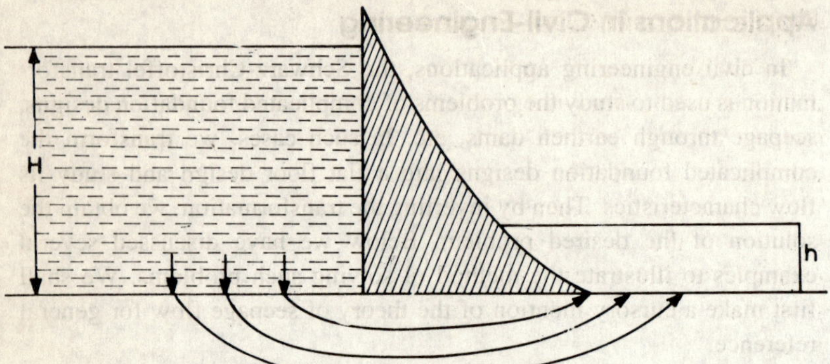

Fig. 35

Seepage may act as a source of weakness for a structure in the following ways:

i) It may cause undermining of the pervious, grannular foundation which would be necessarily followed by the collapse of the whole structure.

ii) The floor or the apron may be forced upwards, owing to the upward pressure of water filtering through the pervious soil under the structure.

The above two sources of weakness caused, respectively, the failure of the old Menufia regulator in Egypt and the Narora weir in India. These are, however, two of the several instances of the consequences of excessive seepage.

For understanding the physical phenomena of seepage, we study the basic nature of the medium and the parameters of seepage flow through it.

Hydraulic Gradient, Seepage Velocity, Discharge Velocity

We now consider an apparatus consisting of a vertical cylinder (Fig. 36) fitted to a certain depth with sand and resting on a meshed screen. Water is led into the cylinder through a tube AB and is let out after filtering through the sand medium through a tap (b). The head required to force the water through the sand filling is measured by means of difference in readings of manometers at C and D. Let H be the head and L be the length of the column of sand then H/L is called the *Hydraulic gradient of filteration* or simply *Hydraulic gradient.*

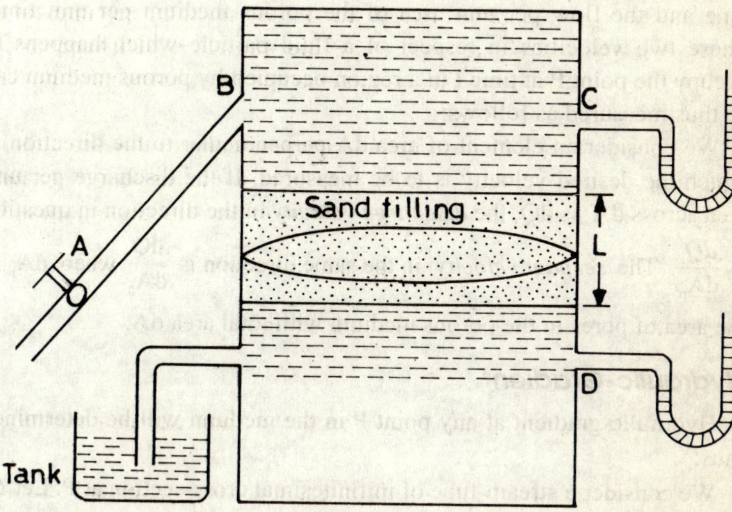

Fig. 36

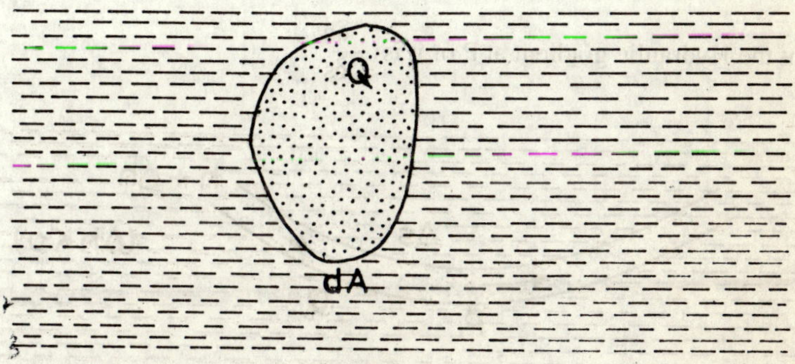

Fig. 37

Now consider a circular cross-section of the cylinderical apparatus. Clearly the normal to it will be the direction of filteration. If A be the area of this section and Q the amount of discharge per unit time across this area, Q/A is called the *discharge velocity* (v). If now A_p denotes the area of the voids (pores) in this cross-section, Q/A_p will be designated as the *seepage velocity* (v_s). Thus, we have formal definitions of seepage velocity and discharge velocity as, respectively, the measure of the quantity of fluid percolating a unit area of the pores (or voids) per unit time and the flow per unit area of the porous medium per unit time. These two velocities in respect of a fluid particle which happens to occupy the point P at time t in a region occupied by porous medium can be thus measured as follows:

We consider an element of area dA perpendicular to the direction in which the desired velocity is to be measured. If the discharge per unit area across dA is dQ, the *discharge velocity* in the direction in question is $\frac{dQ}{dA_p}$. The *seepage velocity* in the same direction is $\frac{dQ}{dA_p}$ where dA_p is the area of pores in the porous medium with total area dA.

Hydraulic-Gradient

Hydraulic-gradient at any point P in the medium will be determined thus:

We consider a stream-tube of inifinitesimal cross-section at P. Let ΔS represent a small length along it beyond P and Δh be the amount of fall in head when the liquid seeps through this distance. $\lim_{\Delta S \to 0} \frac{\Delta h}{\Delta S}$ will then represent the Hydraulic gradient at P or it is $-\frac{\delta h}{\delta s}$

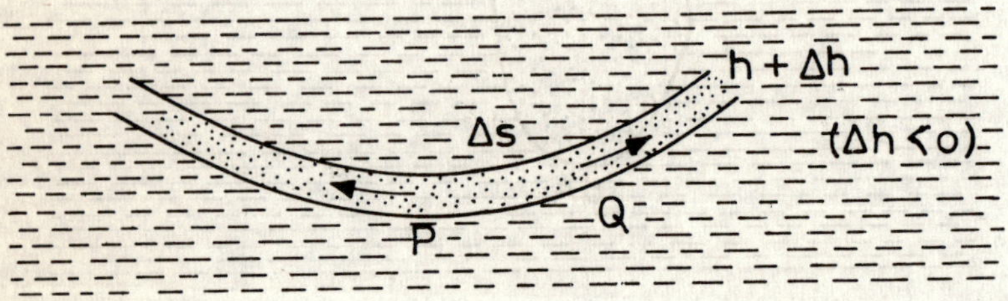

Fig. 38

Darcy's Law

Darcy and Ritter (1856) used the apparatus shown in Fig. 36 and confirmed by experimenting upon various depths of sand filling under different heads that the discharge was proportional to the hydraulic gradient. Now since the discharge is also proportional to the discharge velocity, they established the relation.

$$v = K\left(\frac{H}{L}\right).$$

Here K is called the *Darcy's coefficient* or *coefficient of permeability.* It depends upon the physical properties of the permeable medium but is independent of the discharge velocity.

The general form of Darcy's law expressing the relationship between discharge velocity and hydraulic gradient at a point in space would be

$$v = -K\frac{\partial h}{\partial s}$$

or

$$v = -K \text{ grad } h$$

in vector notation.

It is usual in problems of seepage to refer to the discharge velocity as simply 'velocity', *first*, because it corresponds to the velocity in hydrodynamical problems, and *secondly*, due to the comparative convenience in measuring it. Seepage velocity is not amenable to easy direct measurement.

Prior to the publication of Darcy's relation based on his experiments the formidable nature of seepage flow defied rational analysis. The main contention in the derivation of this relation is that it in no way describes the state of affairs within an individual pore. The medium is, thus, regarded as continuum. The law represents, in a strict sense the statistical equivalent to the Navier-Stoke's equations of motion for the viscous flow of ground water. The limitation on the application of the Navier-stokes equation, *viz.*, that they apply only to a laminar flow is applicable to the Darcy's equation too. As such it is not applicable to almost any of the practical hydraulic problem with which an irrigation or hydraulic engineer may be confronted, the flow there being usually turbulent. The small velocities required for laminar flow in such problems can be realised only in a laboratory experiment. The exception to this is, however, apparent in case of a seepage flow, which is essentially a streamlined laminar motion.

Two Dimensional Problem

Physically all flow systems extend in three dimensions but in many problems the features of the ground water motion are essentially planar. Seepage flow through a wier is, for instance, an example of a two-dimensional flow, the flow in all planes parallel to a breadthwise section being similar due to the length being usually very much longer than the breadth.

If then u, v denote the velocities along x and y axes of the xy-plane in which the representative motion is assumed to take place, we have the Darcy's equations as

$$u = -K\frac{\partial h}{\partial x}, \tag{9.68}$$

$$v = -K\frac{\partial h}{\partial y}. \tag{9.69}$$

Clearly there are three unknowns in the picture u, v and h. We require a third equation to determine them all. This is supplied by assuming that water is incompressible and thus the discharge entering an element ABCD through the faces AB and AC is equal to that escaping through faces CD and BD.

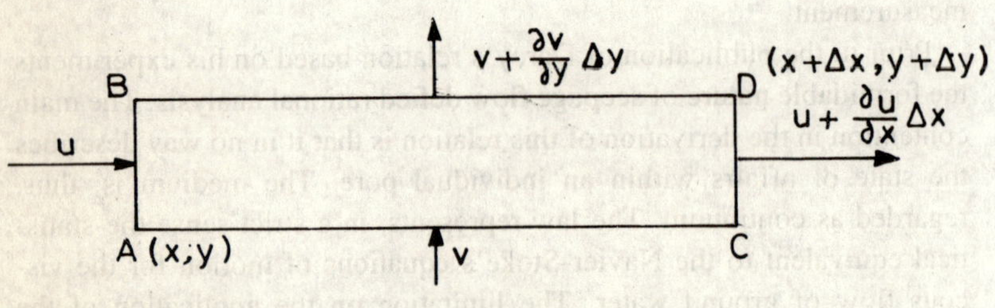

Fig. 39

Discharge entering per unit time is

$$u\,\Delta y + v\Delta x.$$

Discharge leaving per unit time

$$= \left(u + \frac{\partial u}{\partial x}\Delta x\right)\Delta y + \left(v + \frac{\partial v}{\partial y}\Delta y\right)\Delta x.$$

Thus, we have the following equation called the *equation of continuity* as

$$\frac{\partial u}{\partial x} + \frac{\partial v}{\partial y} = 0. \tag{9.70}$$

Stream Function and Stream Lines

We consider a curve in the flow region. The discharge per unit time from left to right across the extent AB of the curve is

$$D = \int_A^B (v \cos\theta - u \sin\theta)\, ds$$

$$= \int_A^B (v\, dx - u\, dy) \tag{9.71}$$

Now since $\frac{\partial u}{\partial x} + \frac{\partial v}{\partial y} = 0$, vdx − udy is an exact differential say $d\psi$, so that (9.71) gives

$$D = \int_A^B d\psi = \psi_B - \psi_A. \tag{9.72}$$

The function ψ is called the *Stream function* of the flow and difference of its value at any two points represents the quantity of discharge. A curve across which there is no flow is called a *stream line*. Thus ψ = const. will be such a line. Equation (9.72) then can be interpreted to give discharge between the stream lines $\psi = \psi_B$ and $\psi = \psi_A$, which is constant throughout the extent of the stream tube thus formed. It is apparent that the velocity will be inversely proportional to the space between the stream-lines due to the discharge remaining constant. Equation ψ = constant can aslo be expressed as

$$\frac{dx}{u} = \frac{dy}{v}$$

or

$$\frac{dy}{dx} = \frac{u}{v}. \tag{9.73}$$

This is then the differential equation of a stream line and suggests also the otherwise apparent fact that the tangent to the stream line gives the direction of the resultant velocity.

Since $d\psi = \frac{\partial \psi}{\partial x} dx + \frac{\partial \psi}{\partial y} dy$

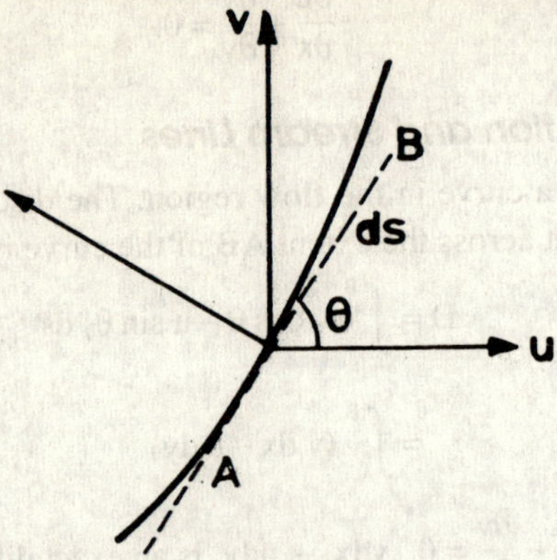

Fig. 40

and $\quad d\psi = vdx - udy,$

we have

$$v = \psi_{,x}\,; \quad u = -\psi_{,y}\,. \tag{9.74}$$

Substituting from here into equation of continuity (9.70), we obtain

$$\psi_{,xx} + \psi_{,yy} = 0, \tag{9.75}$$

which shows that ψ satisfies the Laplace equation and is, thus, a harmonic function.

Velocity Potential Function, Equipotential or Equipressure Lines

The amount of flow per unit time along a curve drawn in the flow region is

$$\int_A^B V \cdot ds = \int_A^B (u \cos\theta + v \sin\theta) ds = \int_A^B udx + vdy. \tag{9.76}$$

If the curve is closed, the integral gives the so-called circulation in the circuit, and is written as

$$\oint (udx + vdy).$$

Considering a point p(x, y) in a region of steady two dimensional flow the circulation in a rectangular circuit around the point is $(v_x - u_y)$ per unit area. This is also called the vorticity at P along perpendicular to the plane of flow.

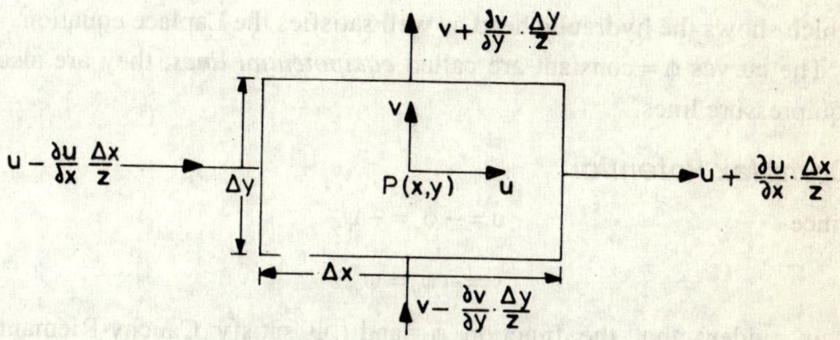

Fig. 41

Thus, we have

$$\oint_C u\,dx + v\,dy = \int_A (v_x - u_y)\,dx\,dy \tag{9.77}$$

where A is the area of region and C its contour.

The motion in a region is called *irrotational* if the vorticity is zero everywhere within it. For such a motion, therefore, we have from (9.77) $v_x - u_y = 0$ everywhere or that $u\,dx + v\,dy$ is an exact differential, say $-d\phi$. Thus for an irrotational motion

$$-d\phi = -\phi_x\,dx - \phi_y dy$$

$$= u\,dx + v\,dy \tag{9.78}$$

or $$u = -\phi_x,\ v = -\phi_y.$$

The function ϕ is called the *velocity potential function*. It apparently exists in case of an irrotational flow while ψ exists for any flow.

In view of the equation of continuity we have

$$\nabla^2\phi = \phi_{,xx} + \phi_{,yy} = 0, \tag{9.79}$$

which shows that ϕ is also a harmonic function. Difference in values of ϕ at two points measures the quantity of flow along any curve joining the points. Also since

$$u = -\phi_{,x} = -Kh_x,\ v = -\phi_{,y} = -Kh_y$$

we have

$$\phi = Kh.$$

Substituting in (9.79), we have

$$h_{,xx} + h_{,yy} = 0 \tag{9.80}$$

which shows the hydraulic head as well satisfies the Laplace equation.

The curves $\phi = \text{constant}$ are called *equipotential lines*; they are also equipressure lines.

Complex Potential

Since
$$u = -\phi_x = -\psi_y$$
$$v = -\phi_y = \psi_x,$$

it is evident that the function ϕ and ψ satisfy Cauchy-Riemann equations. Also that $\phi = \text{constant}$ and $\psi = \text{constant}$ intersect orthogonally. They can be thus treated as the real and imaginary parts of an analytic function of $z = x + iy$.

If $w = \phi + i\psi$ this function w is called the *complex potential* of the problem. Obviously enough $\nabla^2 w = 0$ thus w also satisfies Laplace equation. $\frac{dw}{dz}$ gives the velocity in magnitude and direction. $\frac{dw}{dz}$ is zero or infinite at a point of singularity.

Superposibility of Complex Potentials

If w_1 and w_2 are the complex potentials of two flows, the complex potential of the composite flow will be $w_1 + w_2$. This can be best illustrated with a simple example from Hydrodynamics.

There are two oppositely charged source lines of the same strength situated at points $z = x_1$ and $z = x_2$. The velocity potentials of each of these source lines are, respectively:

$$\phi_1 = -c \ln |z - x_1| \quad \text{and} \quad \phi_2 = c \ln |z_1 - x_2|$$

These are the real parts of the complex potentials :

$$w_1 = -c \ln (z - x_1) \quad \text{and} \quad w_2 = c \ln (z - x_2).$$

The complex potential of the combination of the two source lines is:

$$w = w_1 + w_2 = c \ln [(z - x_1)/(z - x_2)].$$

The equipotential lines are the circles

$$\phi = \text{Re } w = c \ln |z - x_2| / |z - x_1| = \text{constant.}$$

and the lines of force are the orthogonally intersecting lines :

$$\psi = \text{Im } w = c\,(\theta_2 - \theta_1) = \text{constant}$$

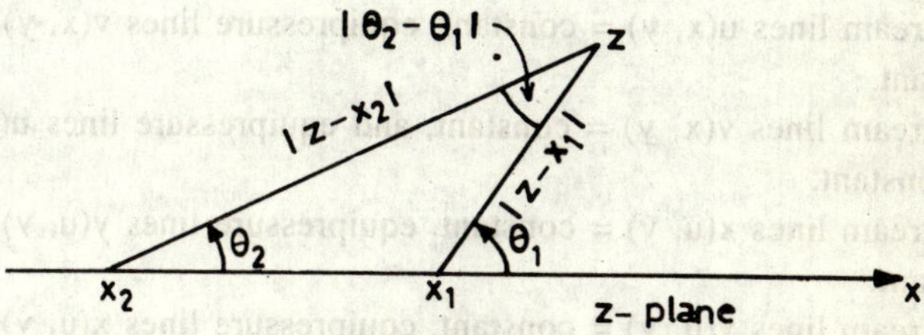

Figure

This function w of z can also be interpreted as complex potential of a Capacitor consisting of two circular cylinders whose axes are parallel but do not coincide. (See figure below)

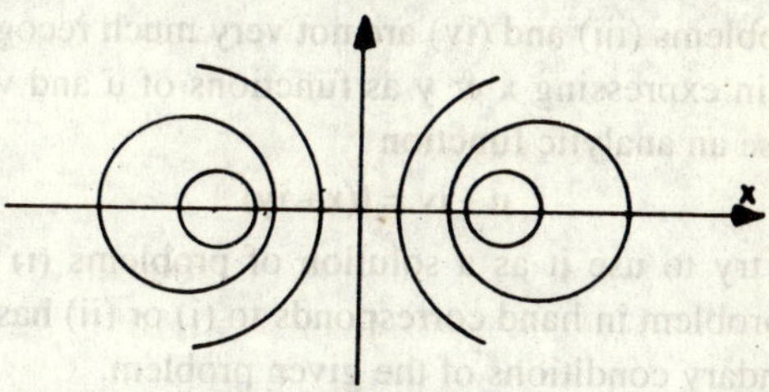

Figure

Through the analytic function or in other words by effecting the conformal transformation through the analytic function

$$w = f(z) \text{ or } \phi + i\psi = \phi(x, y) + i\psi(x, y).$$

the grid of orthogonally intersecting lines ϕ = constant, ψ = constant in the w-zone is mapped into a usually more complicated flow net of orthogonally intersecting lines $\phi(x, y)$ = constant and $\psi(x, y)$ = constant in the w-plane. The obvious advantage of the procedure is that solution of a flow problem in the z-plane can now be easily obtained through the solution of the equivalent problem in w-plane by solving the Laplace equation under the relevent boundary conditions in the w-plane, this being easier due to an easier set of boundary conditions expressible at ϕ = constant, ψ = constant being involved.

Before we proceed further it would be worth while mentioning that any analytic function

$$w = u + iv = f(z) = f(x + iy)$$

provides the solution of the following four problems of irrotational flow:

i) Stream lines u(x, y) = constant, equipressure lines v(x, y) = constant.

ii) Stream lines v(x, y) = constant, and equipressure lines u(x, y) = constant.

iii) Stream lines x(u, v) = constant. equipressure lines y(u, v) = constant.

iv) Stream lines y(u, v) = constant, equipressure lines x(u, v) = constant.

The complex potential in the above cases would be, respectively

i) w = f(z), (ii) iw = if(z), (iii) z = F(w), (iv) iz = iF(w),

where z = F(w) is derived from w = f(z).

Usually problems (iii) and (iv) are not very much recognizable due to complication in expressing x & y as functions of u and v. Thus, whenever we choose an analytic function

$$u + iv = f(x + iy)$$

we generally try to use it as a solution of problems (i) and (ii). Thus whether our problem in hand corresponds to (i) or (ii) has to be guessed from the boundary conditions of the given problem.

The usual procedure is to consider different analytic functions and analyse as to which problems are solvable through them. The main thing to be remembered however is that body is one of the streamlines and this helps the analysis.

We consider the analytic function

$$w = u + iv = (x + iy)^2, \tag{9.81}$$

so that

$$u = x^2 - y^2,$$

$$v = 2xy.$$

Since u = constant and v = constant are two types of hyperbolas, the flow net will be as in Fig. 42.

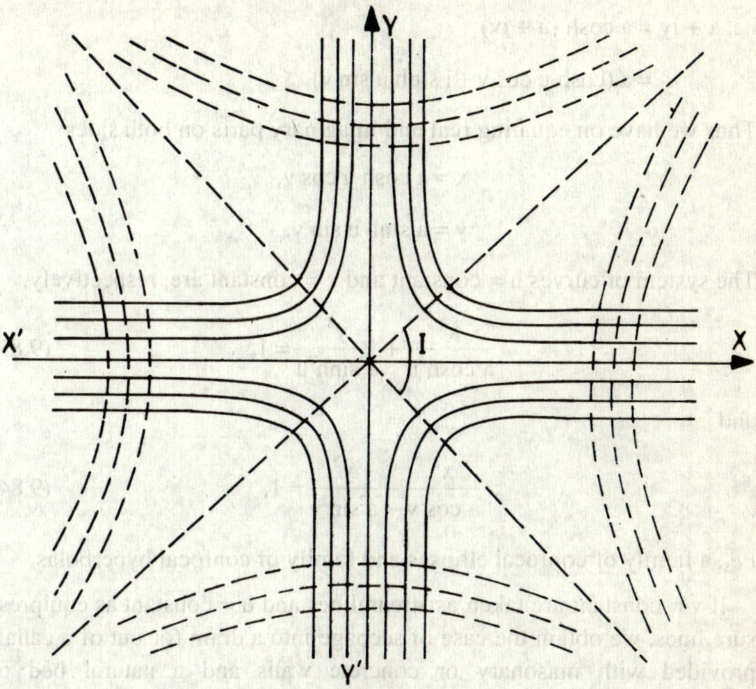

Fig. 42

The thick lines being v = constant and the dotted lines u = constant

If now the lines u = constant are chosen as stream lines and v = constant the equipressure lines, we can, for instance, solve the problem of flow in a corner like I. With the choice of u = constant as the equipotential lines and v = constant as the stream lines the problem of flow through a symmetrical embankment is solved, the reason being that by symmetry the streamline in the middle will go down and bifurcate into two. Its equation is xy = 0 which in a way suggests the solution.

As a second example, consider the transformation

$$w = \cosh^{-1}(z/a)$$

or $$z = a \cosh w. \tag{9.82}$$

i.e. $x + iy = a \cosh (u + iv)$

$$= a (\cosh u \cos v + i \sinh u \sin v).$$

Thus we have on equating real and imaginary parts on both sides

$$x = a \cosh u \cos v,$$

$$y = a \sinh u \sin v.$$

The system of curves u = constant and v = constant are, respectively,

$$\frac{x^2}{a^2\cosh^2 u} + \frac{y^2}{a^2\sinh^2 u} = 1, \tag{9.83}$$

and

$$\frac{x^2}{a^2\cos^2 v} - \frac{y^2}{a^2\sin^2 v} = 1, \tag{9.84}$$

i.e., a family of confocal ellipses and family of confocal hyperbolas.

If v = constant are taken as streamlines and u = constant as equipressure lines, we obtain the case of seepage into a drain (or out of a canal) provided with masonary or concrete walls and a natural bed of permeable material (Fig. 43).

If, on the other hand, ellipses are the stream lines and the hyperbolas considered as the equipressure lines, we get a very important and practically significant case, *viz.*, that of a flat floor beneath a weir (Fig. 44).

The problem of seepage below a weir on a flat floor AB can also be posed as follows:

Show that the function

$$z = a \cosh w = a \cosh (u + iv) = a \cos(iu - v)$$

can be fitted to the solution of the problem of seepage flow below a weir on a flat floor by choosing u and v as stream function and velocity potential, respectively. Also find the total uplift pressure on the floor and moment of uplift forces about one of its edges.

Let us take the floor to be AB along the x-axis, y-axis being taken along the downward normal to the floor and the origin being at the mid-point of the floor.

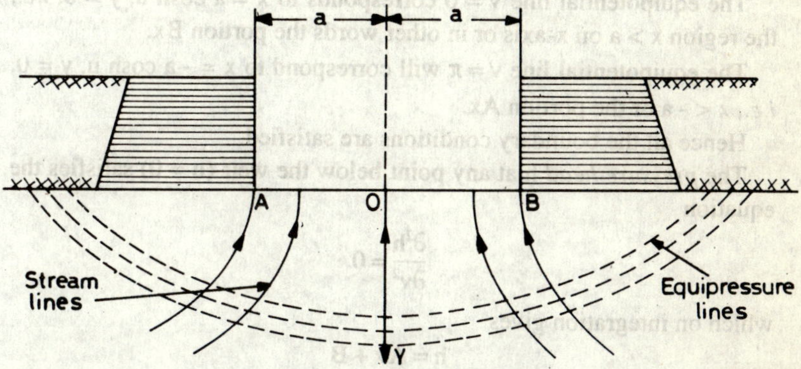

Fig. 43

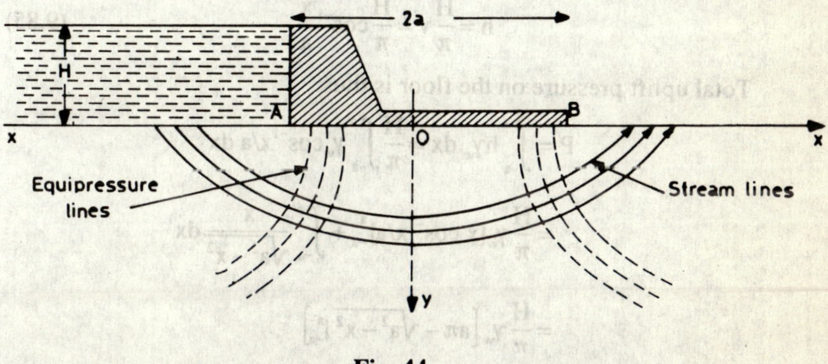

Fig. 44

The boundary conditions of the problem, then are:

i) The portion AB of the line $y = 0$ is a stream line;

ii) The portion Ax of the line is an equipressure line, and

iii) the portion Bx of the line $y = 0$ is another equipotential line.

The stream line $u = 0$ is

$x = a \cos v, y = 0$ or $-a \le x \le a$

i.e., it corresponds to the flat floor AB if the floor length is 2a.

The equipotential line $v = 0$ corresponds to $x = a \cosh u, y = 0$, *i.e.*, the region $x > a$ on x-axis or in other words the portion Bx.

The equipotential line $v = \pi$ will correspond to $x = -a \cosh u, y = 0$, *i.e.*, $x < -a$ or the portion Ax.

Hence all the boundary conditions are satisfied.

The *pressure head* h at any point below the weir (u = 0) satisfies the equation

$$\frac{\partial^2 h}{\partial v^2} = 0$$

which on integration gives

$$h = Av + B$$

where A and B are to be determined from the boundary conditions.

$$h = H, v = \pi,$$

$$h = 0, v = 0.$$

Applying these conditions we have $h = H/\pi$ and $B = 0$, so that

$$h = \frac{H}{\pi} v = \frac{H}{\pi} \cos^{-1} \frac{x}{a}. \tag{9.85}$$

Total uplift pressure on the floor is thus

$$P = \int_{-a}^{a} h\gamma_w \, dx = \frac{H}{\pi} \int_{-a}^{a} \gamma_w \cos^{-1} x/a \, dx$$

$$= \frac{H}{\pi} \gamma_w [x \cos^{-1} x/a]_{-a}^{a} + \int_{-a}^{a} \frac{x}{\sqrt{a^2 - x^2}} dx$$

$$= \frac{H}{\pi} \gamma_w \left[a\pi - \sqrt{a^2 - x^2} \Big|_{-a}^{a} \right]$$

$$= a \gamma_w H, \tag{9.86}$$

where γ_w is unit weight of fluid.

Total moment of uplift forces on the floor about an edge is

$$M = \int_{-a}^{a} h\,\gamma_w (x + a)\, dx$$

$$= \frac{H}{\pi}\gamma_w \int_{-a}^{a} (x + a) \cos^{-1} x/a\, dx$$

$$= \frac{3}{4} H \quad \gamma_w a^2 \tag{9.87}$$

Note: If the pressure head upstream is H and downstream is h, it can be verified that

$$p = a\gamma_w\,(h + H), M = \frac{\gamma_{w^{a^2}}}{4}(5h + 3H). \tag{9.88}$$

See Figure 45 below:

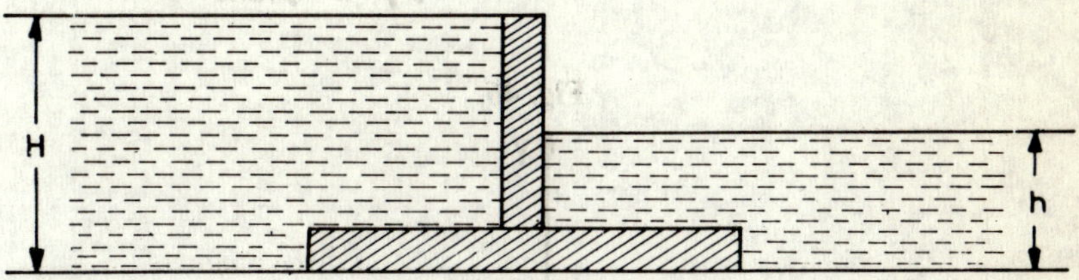

Fig. 45

Example 10. The Transformation Formula

Let us be given a polygon of the type shown in Fig. 46. The open polygon has the contour AEDCB. The interior is the region on the left of an observer treading the boundary from B to C, C to D, D to E, E to A with his head towards the reader. Interior angles are marked in the figure. This figure can be looked upon as representing a complicated foundation design.

The figure in w-plane transforms onto the z-plane in the following way:

Origin is placed between E′, C′, the points corresponding to E.C. D′ the point corresponding to D such that OD′ : OC′ = k. The points E′ and C′ are taken to have the coordinates (–1, 0) and (1, 0) in z-plane.

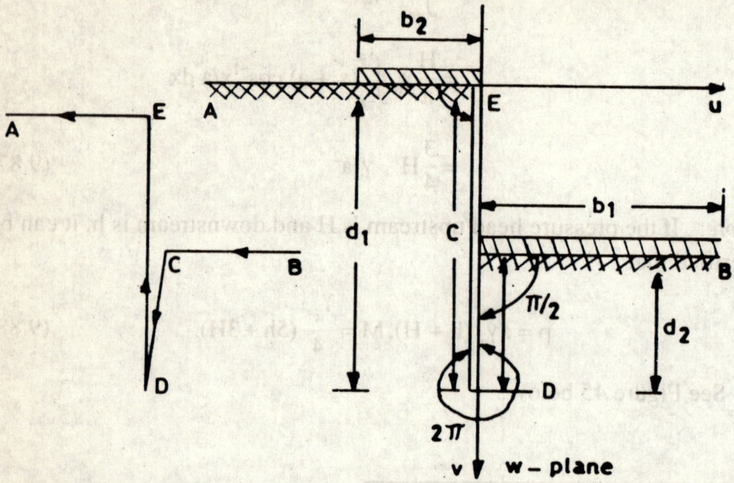

Fig. 46

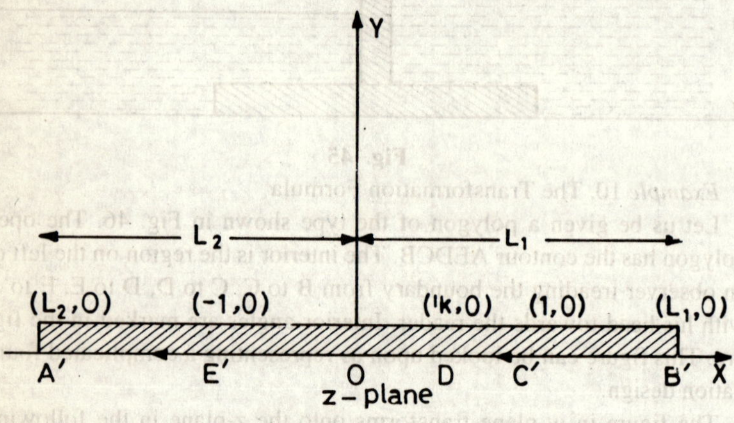

Fig. 47

Since $z_1 = -1, z_2 = k, z_3 = 1,$

and $\alpha_1 = \pi/2, \alpha_2 = 2\pi, \alpha_3 = \pi/2,$

the appropriate Schwarz-Christoffel transformation would be given by

$$\frac{dw}{dz} = K(z^2 - 1)^{-1/2}(z - k)$$

or on integration

$$w = K\sqrt{z^2 - 1} - K\,k\log(z + \sqrt{z^2 - 1}) + A. \tag{9.89}$$

where A, K and k are to be determined.

The conditions connecting w and z are

$$z = 0 \text{ when } w = \frac{i(d_1 - d_2)}{2},$$

$$z = -1, \text{ when } w = 0$$

$$z = 1, \text{ when } w = i(d_1 - d_2)$$

so that we have

$$\frac{i(d_1 - d_2)}{2} = Ki - Kk\log i + A$$

$$0 = A - Kk\log(-1)$$

$$i(d_1 - d_2) = A - Kk\log(1), \tag{9.90}$$

providing

$$A = i(d_1 - d_2), Kk = \frac{d_1 - d_2}{\pi} \tag{9.91}$$

Also on substituting the condition

$z = k$ when $w = id_1$ in Equation (9.89) one has

$$id_1 = K\sqrt{k^2 - 1} - Kk\log(k + \sqrt{k^2 - 1}) + i(d_1 - d_2)$$

$$= \frac{d_1 - d_2}{\pi k}\sqrt{k^2 - 1} - \frac{d_1 - d_2}{\pi}\log(k + \sqrt{k^2 - 1})$$

$$+ i(d_1 - d_2). \tag{9.92}$$

since $k < 1$, we choose $k = \cos\theta$, so that the above relation gives after simplification

$$\delta = \frac{d_2}{d_1 - d_2} = \frac{\tan\theta - \theta}{\pi}. \tag{9.93}$$

Khosla has tabulated values of k for different values of δ on the basis of above relation.

Now since the points A′ and B′ are taken as $(-L_2, 0)$; $(L_1, 0)$ we have the following conditions:

$$z = -L_2, w = -b_2$$

$$z = L_1, w = b_1 + i(d_1 - d_2)$$

which when substituted in (9.89) give

$$-b_2 = \frac{d_1 - d_2}{\pi k}\sqrt{L_2^2 - 1} - \frac{d_1 - d_2}{\pi}\log\left(-L_2 + \sqrt{L_2^2 - 1}\right) + i(d_1 - d_2) \qquad (9.94)$$

and

$$b_1 + i(d_1 - d_2) = \frac{d_1 - d_2}{\pi k}\sqrt{L_1^2 - 1} - \frac{d_1 - a_2}{\pi} + \log\left(-L_1 + \sqrt{L_1^2 - 1}\right)$$

$$+ i(d_1 - d_2). \qquad (9.95)$$

We substitute $L_1 = \cosh\gamma_1$, $L_2 = \cosh\gamma_2$, so that (9.94), (9.95) provide

$$-b_1 = \frac{d_1 - d_2}{\pi k}\sinh\gamma_1 - \frac{d_1 - d_2}{\pi}\log(-\cosh\gamma_2 + \sinh\gamma_2)$$

and

$$b_1 = \frac{d_1 - d_2}{\pi K}\sinh\gamma_1 - \frac{d_1 - d_2}{\pi}\log(\cosh\gamma_1 + \sinh\gamma_1)$$

or

$$-\pi k\delta_2 = \sinh\gamma_2 + k\,\gamma_2 \qquad (9.96)$$

$$\pi k\delta_1 = \sinh\gamma_1 - k\,\gamma_1 \qquad (9.97)$$

where

$$\delta_1 = \frac{b_1}{d_1 - d_2}, \; \delta_2 = \frac{b_2}{d_1 - d_2}. \qquad (9.98)$$

Thus the problem of the complicated foundation transforms to that of a flat foundation design. One can, therefore, study its flow characteristics on the lines of the flat floor problem discussed earlier.

---

*Design of weirs on permeable foundations by Rai Bahadour A.N. Khosla, Dr. N.K. Bose and Dr. E. Mckenzie Taylor, Publication N12 of the Central Board of Irrigation, *India*, (Simla 1936).

The complex potential Ω for the flow will be related to the variable z as follows:

$$z+\frac{L_1-L_2}{2}=\frac{L_1+L_2}{2}\cosh\Omega \qquad (9.99)$$

(We can note that the only difference in flat floor problem discussed in theory of seepage flow was that the origin there was at the middle point of the floor).

or we have

$$z=\lambda\cosh\Omega-\lambda_1$$

where

$$\lambda_1=\frac{L_1-L_2}{2} \quad \text{and} \quad \lambda=\frac{L_1+L_2}{2}$$

so that the relation between w and Ω at any point is given by

$$w=\frac{d_1-d_2}{\pi k}\sqrt{(\lambda\cosh\Omega-\lambda_1)^2-1}$$

$$-\frac{d_1-d_2}{\pi z}\log\left[(\lambda\cosh\Omega-\lambda_1)+\sqrt{(\lambda\cosh\Omega-\lambda_1)^2-1}\right]+i(d_1-d_2). \qquad (9.100)$$

The value of the uplift pressure at any point of the zigzag polygon can be found from the above relation by finding the value of ϕ at the point. Any point on the upper floor AE is (x, 0) while on the lower floor CB is $(x, d_1 - d_2)$. Any point on the sheet pilings ED or DC is (0, y).

From the general case discussed above, the following important cases are derivable:

Case(i) Apron provided with piling at the downstream end, with a step behind it.

This design is derivable from the general design by putting b_1, the length of the downstream floor zero.

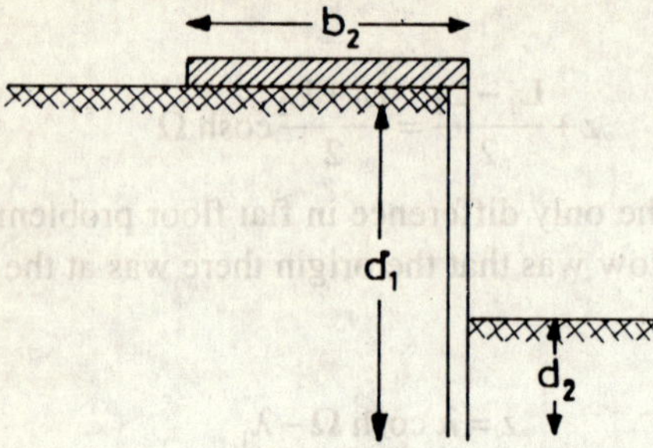

Fig. 48

Case (ii) Single pile-line with fall ($b_1 = b_2 = 0$)

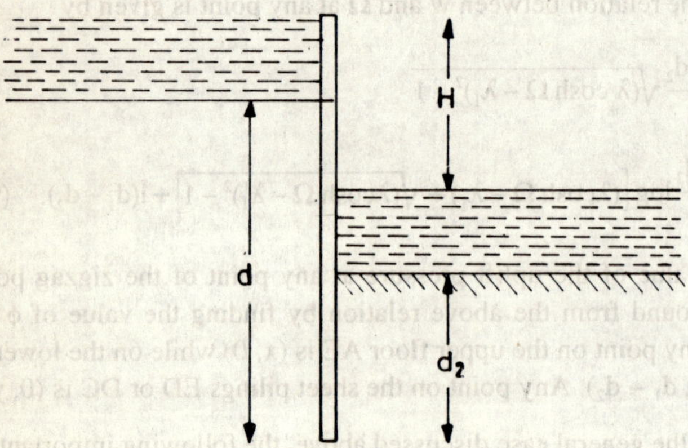

Fig. 49

Case (iii) Single pile-line. Equal fill on either side. No apron.

Due to some peculiarities we choose to study this case fully. Since here $d_1 = d_2$, we have from (9.93) $\tan\theta \to \infty$ or $\theta = \pi/2$, which in its turn implies $k = 0$.

Now $k = 0$, $d_1 = d_2$ gives rise to a term of the form (0/0), *i.e.*, an indeterminate term in the right hand side of (9.94).

Considering then (9.93) in the following form

$$\sin\theta - \theta\cos\theta = \pi d_2 \frac{k}{d_1 - d_2},$$

we have when $k = 0$, $d_1 = d_2 = d$ (say),

$$1 - 0 = \pi d_2 \cdot \lim_{d_1 \to d_2} \left[\frac{k}{d_1 - d_2}\right]$$

which provides

$$\lim_{d_1 \to d_2} \left[\frac{k}{d_1 - d_2}\right] = \frac{1}{\pi d}. \tag{9.101}$$

Using (9.101), equation (9.100) gives

$$w = d\sqrt{z^2 - 1}, \tag{9.102}$$

which provides the appropriate Schwarz-Christoffel transformation in this case.

The map of Fig. 50 in the z-plane is the following figure.

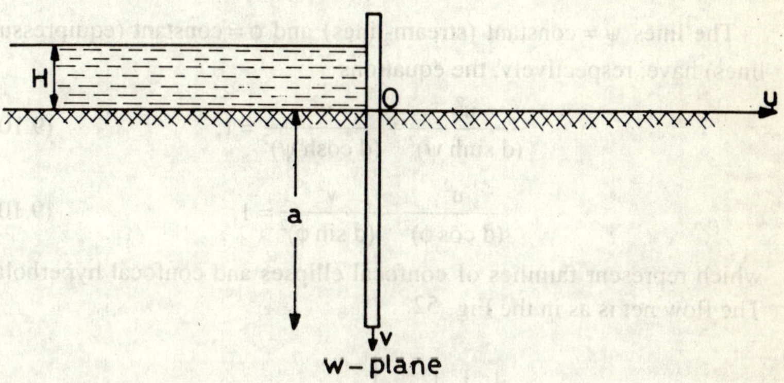

Fig. 50

The complex potential Ω for the flat floor problem is given by

$$z = \cos \Omega. \tag{9.103}$$

From (9.102) and (9.103) we have the relation connecting w and Ω as follows:

$$w = id \sin \Omega \tag{9.104}$$

so that $\quad u + iv = id \sin(\phi + i\psi)$

$$= id \{\sin\phi \cosh\psi + i\cos\phi \sinh\psi\}$$

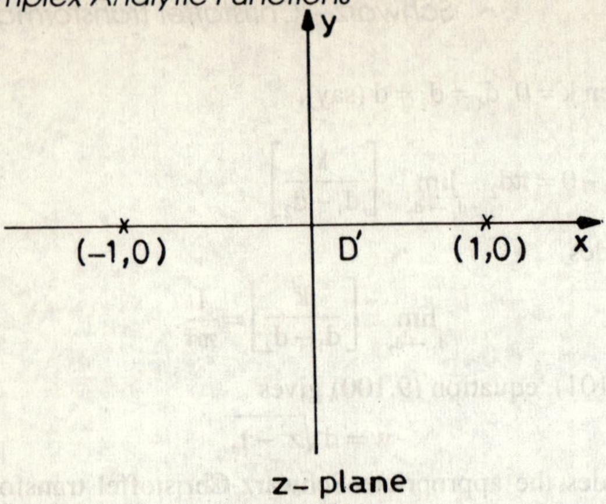

Fig. 51.

or we have

$$u = -d \cos\phi \sinh\psi$$

$$v = d \sin\phi \cosh \psi. \qquad (9.105)$$

The lines ψ = constant (stream-lines) and ϕ = constant (equipressure lines) have, respectively, the equations

$$\frac{u^2}{(d \sinh \psi)^2} + \frac{v^2}{(d \cosh \psi)^2} = 1, \qquad (9.106)$$

$$\frac{u^2}{(d \cos \phi)^2} - \frac{v^2}{(d \sin \phi)^2} = 1, \qquad (9.107)$$

which represent families of confocal ellipses and confocal hyperbolas. The flow net is as in the Fig. 52.

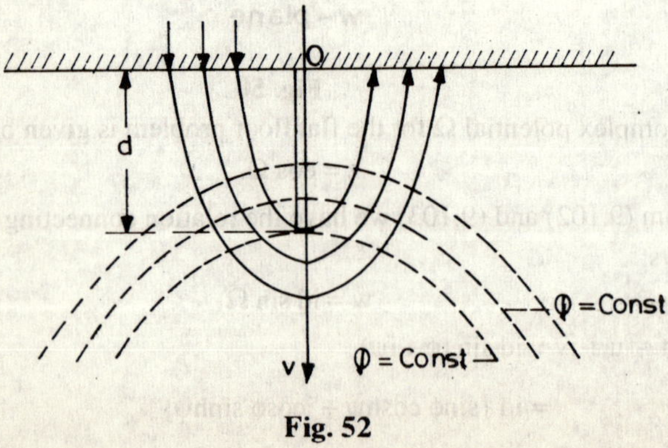

Fig. 52

The equipressure line $\phi = 0$ is $x = -d \sinh \psi$, $y = 0$, *i.e.*, the entire negative x-axis. The stream-line $\psi = 0$ is $x = 0$, $y = d \sin \phi$, *i.e.*, the pile itself.

Pressure at Any Point on the Pile

Any point on the pile is $w = iv$. Thus we have trom (9.105)

$$v = d \sin \phi. \tag{9.108}$$

Since the total head H corresponds to π, we have

$$\phi = \pi \frac{P}{H}$$

where P is he pressure at any point on the pile or thus

$$P = \frac{H}{\pi} \phi = \frac{H}{\pi} \sin^{-1}\left(\frac{v}{d}\right). \tag{9.109}$$

Solution of problems concerning quite a few other foundation designs, *e.g.*, those of a floor with pile line at one end, no step, or with pile line at one end are derivable from the general solution (9.100) for the general design

Example 11. Seepage through an Earthen Dam

We consider a dam of width 2b founded at the surface of an infinite permeable medium (Fig. 53) in z-plane and a polygon $A_\infty BCD_\infty$ with BC = 2a and its vertices A and D at infinity (Fig. 54) in w-plane.

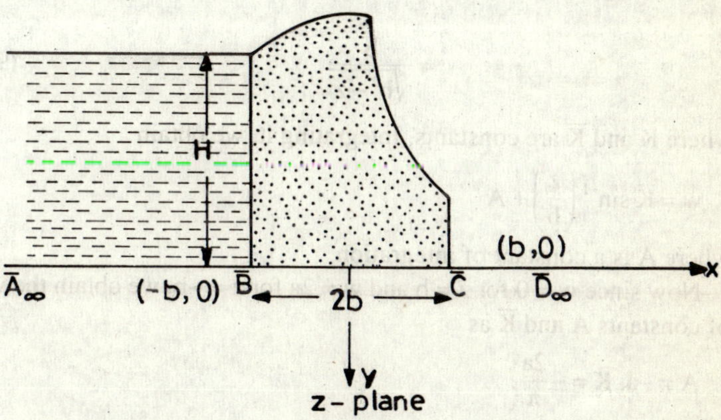

Fig. 53.

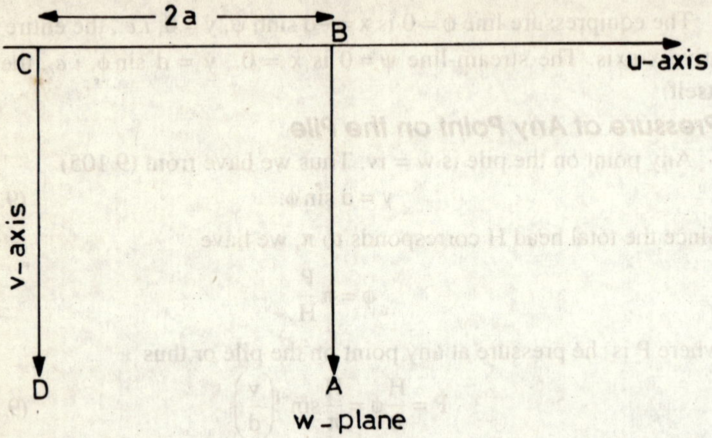

Fig. 54.

We now apply the Schwarz-Christoffel transformation to transform the interior of polygon $A_\infty BCD_\infty$ into the upper-half of z-plane in such a manner that the vertices at A_∞, D_∞ correspond to $\bar{A}_\infty, \bar{D}_\infty$ in z-plane and the points B and C transform into the points (–b, 0) and (b, 0), respectively, in z-plane. The Schwarz-Christoffel transformation then gives.

$$\frac{dw}{dz} = K(z-b)^{-1/2}(z+b)^{-1/2}$$

$$= \frac{\bar{K}}{\sqrt{b^2 - z^2}} \quad (9.110)$$

where K and $\bar{K}$ are constants. Integrating it, we obtain

$$w = \bar{K} \sin^{-1}\left(\frac{z}{b}\right) + A$$

where A is a constant of integration.

Now since w = 0 for z = b and w = 2a for z = –b, we obtain the values of constants A and $\bar{K}$ as

$$A = -a, \bar{K} = -\frac{2a}{\pi}.$$

Hence the solution becomes

$$w = -a \left[\frac{2}{\pi} \sin^{-1}\left(\frac{z}{b}\right) + 1\right]$$

$$= \frac{2a}{\pi}\cos^{-1}\left(\frac{z}{b}\right)$$

or

$$z = b\cos\left(\frac{\pi w}{2a}\right) \tag{9.111}$$

i.e., $x + iy = b\cos\left[\frac{\pi(u + iv)}{2a}\right]$

Equating real and imaginary parts on both sides, we obtain

$$x = b\cos\frac{\pi u}{2a}\cosh\frac{\pi v}{2a}, \tag{9.112}$$

$$y = b\sin\frac{\pi u}{2a}\sinh\frac{\pi v}{2a}. \tag{9.113}$$

The u and v eliminants of (9.112) and (9.113) are

$$\frac{x^2}{b^2\cosh^2\left(\frac{\pi v}{2a}\right)} + \frac{y^2}{b^2\sinh^2\left(\frac{\pi v}{2a}\right)} = 1, \tag{9.114}$$

$$\frac{x^2}{b^2\cos^2\left(\frac{\pi u}{2a}\right)} - \frac{y^2}{b^2\sin^2\left(\frac{\pi u}{2a}\right)} = 1, \tag{9.115}$$

respectively, Equation (9.114) represents a family of ellipses, in which any one ellipse is obtained by choosing a value of v; thus each ellipse is a flow-line. Equation (9.115) is the equation of a family of hyperbolas conjugate to the ellipses, and having a common focus ($\pm b, 0$). It gives the equipotential lines where any one equipotential is obtained by choosing a value of u and inserting it in equation.

To determine the constant a, we observe in Fig. 52 that the head at point $\overline{B}$ is H. Also for this point v is zero.

Thus putting v = 0 in (9.112), we have

$$\frac{\pi u}{2a} = \cos^{-1}\left(\frac{x}{b}\right) \tag{9.116}$$

It determines the distribution of head along the base of the dam. Now at $\overline{B}$ x = −b and u = H so that 2a = H, and Equations (9.114), (9.115) take the forms

$$\frac{x^2}{b^2\cosh^2\left(\frac{\pi v}{H}\right)} + \frac{y^2}{b^2\sinh^2\left(\frac{\pi v}{H}\right)} = 1, \tag{9.117}$$

$$\frac{x^2}{b^2\cos^2\left(\frac{\pi u}{H}\right)}-\frac{y^2}{b^2\sin^2\left(\frac{\pi u}{H}\right)}=1. \quad (9.118)$$

The head, (9.116), varies as a cosine function along the base of the dam with theoretical gradients of infinity at both heel and toe. Since at the heel the direction of flow is downward, this gradient is of little consequence there, but its upward direction at the toe is a potentially dangerous condition. In practice, dams or aprons are never constructed on the surface of the ground in this way, in part because of these high toe gradients.

Two families of hyperbolas and ellipses have been constructed in Fig. 55. The ellipses of flow approach the form of circles with increasing distance from the dam foundation, and the hyperbolas of total head tend toward radial straight lines.

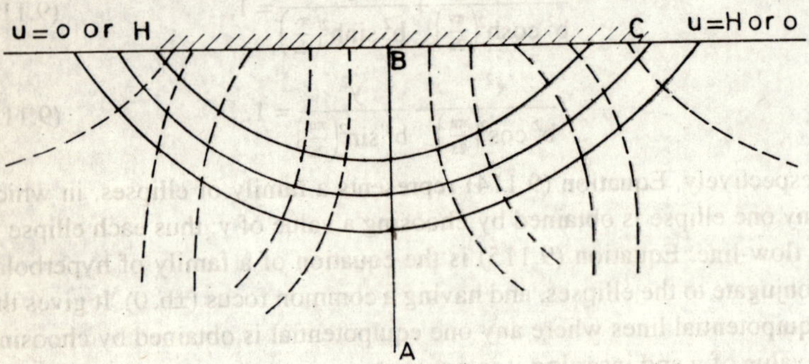

Fig. 55

If instead of the system described above, we consider the conjugate case, where ellipses represent equipotentials and the hyperbolas describe flow lines, we can see that this new system describes flow coming from a strip source of constant head along the position of the base of dam, and the upstream and downstream horizontal boundaries which were equipotentials for the dam become impervious boundaries (or lines of symmetry).

Before we conclude, let us also consider the transformation which should be used for *Mapping A Strip on to A Circle*.

Let the strip shown in the w-plane have width 2a and it spreads infinitely from A_∞ to B_∞. The points A_∞ and B_∞ map into points A and B of

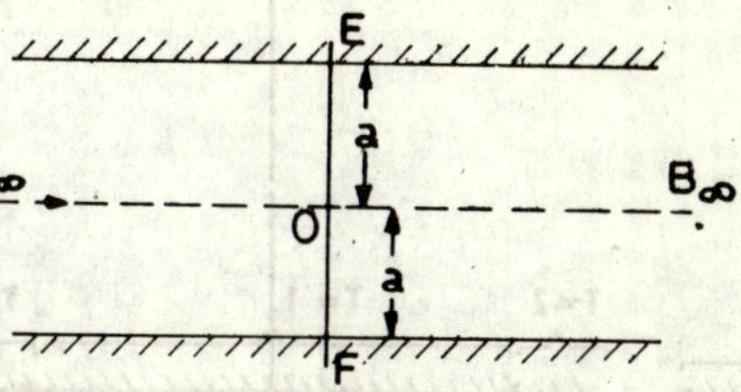

w - plane

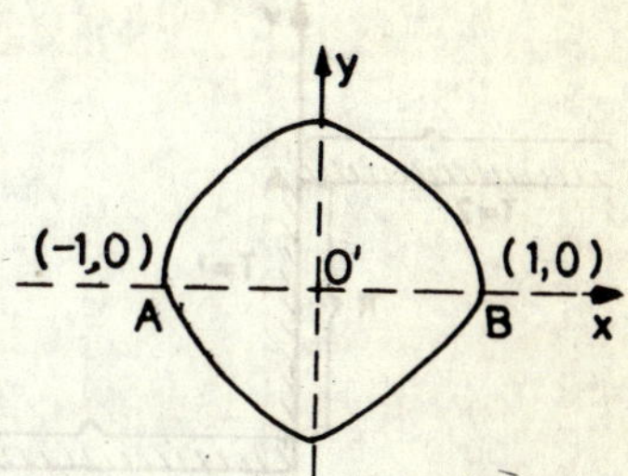

z - plane

the z-plane having positions $x_1 = -1$ and $x_2 = 1$, so that the point $w = 0$ in the w-plane maps into the point $z = 0$ in the z-plane. Since the angles at A_∞ and B_∞ are zero, the equation (9.1) of transformation takes the form:

$$dw/dz = K\,(z-1)^{\frac{0}{\pi}-1}\,(z+1)^{\frac{0}{\pi}-1}$$

$$= K/(z^2-1)$$

or integrating we get :

$$w = (K/2)\log[(1-z)/(1+z)] + C, \text{ where C is a constant.}$$

Now since $z = 0$ corresponds to $w = 0$ and $z = i$ to $w = ia$, we have $C = 0$, also:

$$ia = (K/2)\log[(1-i)/(1+i)]$$

$$= (K/2)\log(\exp(-i\pi/4)/\exp(i\pi/4))$$

$$= -iK(\pi/4)$$

or $\qquad K = -4a/\pi.$

Thus $w = (2a/\pi)\log[(1+z)/(1-z)]$, and $z = \tanh(\pi w/4a)$.

It is easily derivable from above that $w = ia$ corresponds to $z = -i$.

EXERCISE

1. By first mapping into the z-plane as indicated, find the steady-state temperature at any point in the region shown in the following figure:
   Solution:

We could take the figure in the w-plane as a polygon whose vertices

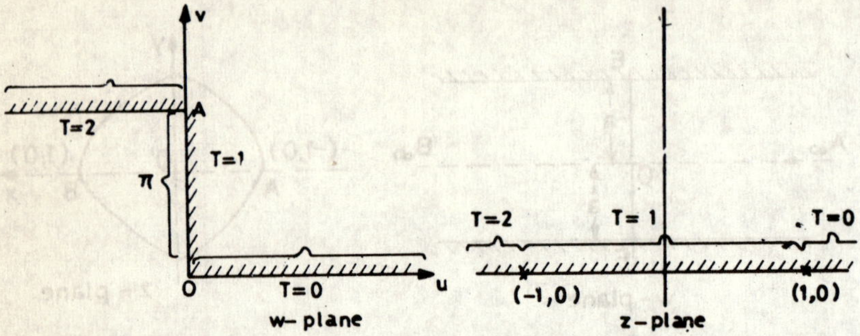

A and O are in the finite plane, the internal angles at these points are $3\pi/2$ and $\pi/2$ and they transform to $z = -1$ and $z = +1$ in the z-plane. Thus, the Schwarz-Christoffel transformation for this case is governed by the differential equation:

$$dw/dz = K\,(z+1)^{1/2}\,(z-1)^{-1/2}$$

$$= K\sqrt{(z+1)/(z-1)}$$

$$= K\,(z/\sqrt{z^2-1}) + (K/\sqrt{z^2-1})$$

or integrating:

$$w = K\,(\sqrt{z^2-1} + \cosh^{-1} z) + L.$$

Now since $z = -1$ corresponds to $w = i\pi$ and $z = +1$ corresponds to $w = 0$, $L = 0$ and $K = 1$, so that the transformation can be written as:

$$w = (\sqrt{z^2-1} + \cosh^{-1} z)$$

Now the steady state Temperature T satisfies the equation $\dfrac{\partial^2 T}{\partial u^2} + \dfrac{\partial^2 T}{\partial v^2} = 0$, which transforms to $\dfrac{\partial^2 T}{\partial x^2} + \dfrac{\partial^2 T}{\partial y^2} = 0$ in the z-plane. The functions $\phi_1 = \tan^{-1}\dfrac{y}{x+1}$ and $\phi_2 = \tan^{-1}\dfrac{y}{x-1}$ are the imaginary parts of two analytic functions and thus are harmonic. A solution of $\nabla^2 T = 0$ in a two dimensional z-plane can, thus, be taken as $T = \frac{1}{\pi}[\phi_1 + \phi_2] = \frac{1}{\pi}\tan^{-1}[2xy/x^2 - y^2 - 1]$. It can be easily seen that this expression for T satisfies the boundary conditions:

$$T = 2, x < -1$$

$$= 1, -1 < x < 1$$

$$= 0, x > 1$$

2. Find the transformation which will map the region in w-plane into the upper half of z-plane as indicated.

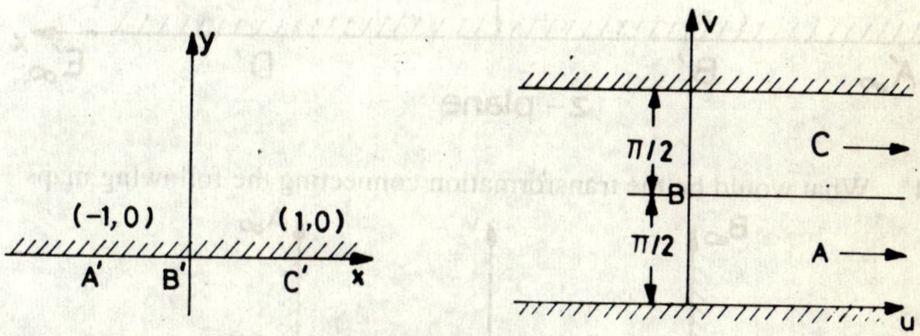

w-plane and z-plane

3. Propose a proper Schwarz-Christoffel transformation for the mapping of the region with a slit of height h in w-plane onto the upper half of z-plane as indicated in the following figure.

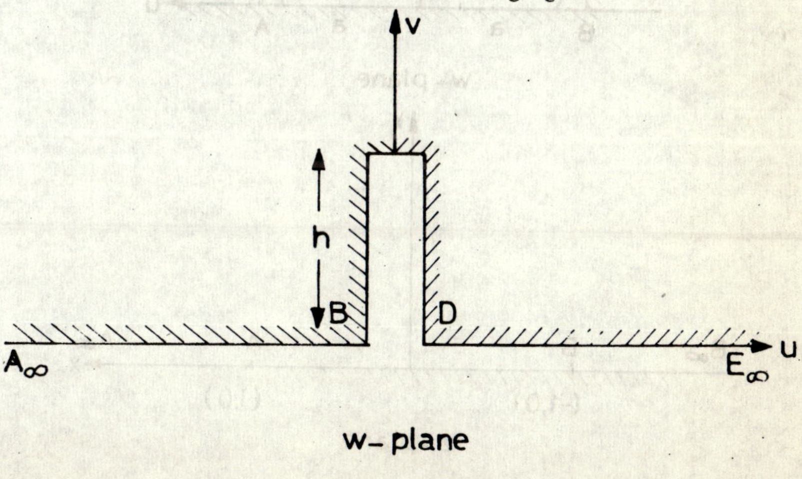

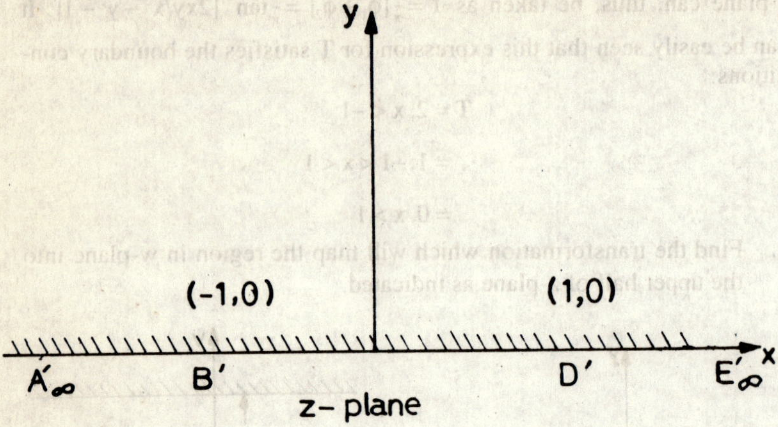

4. What would be the transformation connecting the following maps:

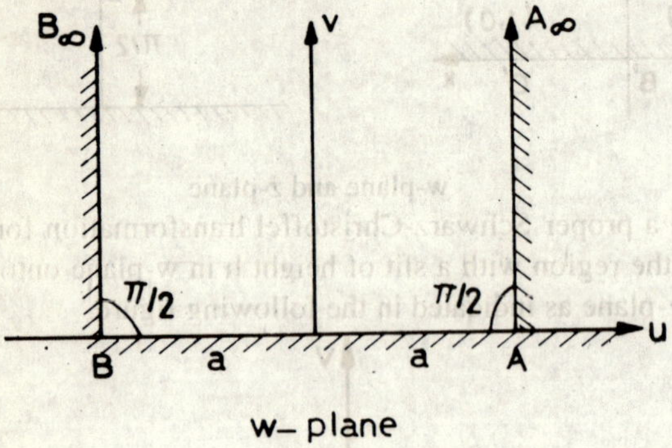

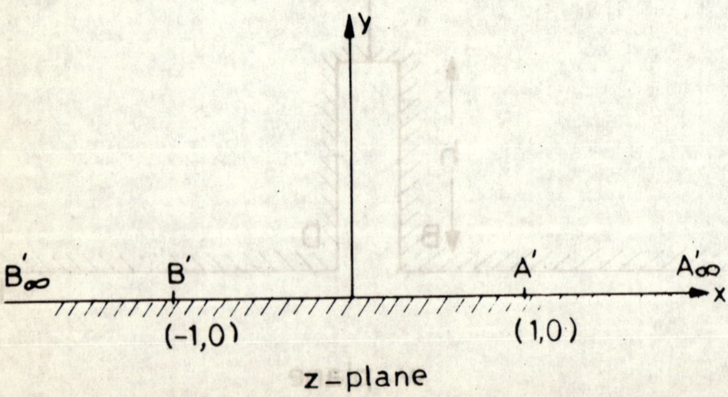

5. An infinite channel with parallel sides at a distance 1 apart has a narrow slit at a point 0. A perfect fluid flows through the slit at a constant rate R per unit length of the slit. Find the complex potential. Deduce also the complex potential for the related problems:
   i) A source equidistant from two-plane sides of an infinite channel.
   ii) A source at one vertex of a semi-infinite rectangular channel.
6. Find the complex potential for the horizontal flow, which is uniform at infinity with speed U, of a perfect fluid over a step of height h in the otherwise horizontal bed of a semi-infinite ocean.

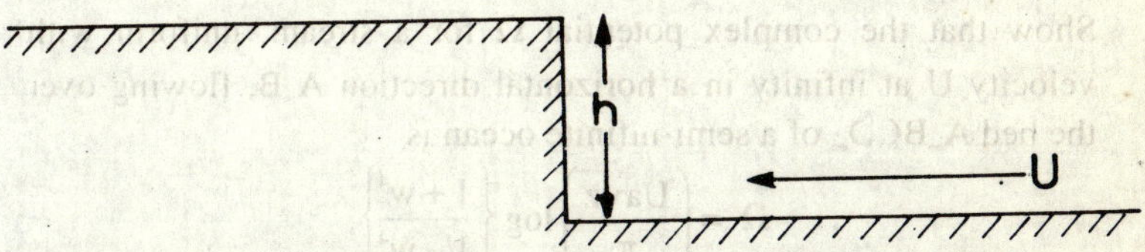

7. Find the complex-potential for a uniform flow in a channel whose sides are parallel planes and which has a sudden discontinuity of section.
8. A semi-infinite rectangular electrode (at potential V_0) has a thin finite projection (length L) at the middle of, and perpendicular to, its shortest side (length 2a). A semi-infinite thin plate at zero potential is placed in line with this projection and at a distance b from it. Show that the complex potential Ω can be found from

$$z = \left(\frac{2a}{\pi}\right)\{w^{1/2}(1+w)^{1/2} + \sinh^{-1} w^{1/2}\},$$

$$\Omega = \left(\frac{2V_0}{\pi}\right)\log\left\{\frac{(w-\lambda)^{1/2} + (w-\mu)^{1/2}}{(\mu-\lambda)^{1/2}}\right\},$$

where λ and μ are determined from

$$\lambda^{1/2}(1+\lambda)^{1/2} + \sinh^{-1}\lambda^{1/2} = \frac{\pi L}{2a};$$

$$\mu^{1/2}(1+\mu)^{1/2} + \sinh^{-1}\mu^{1/2} = \frac{\pi(L+b)}{2a}.$$

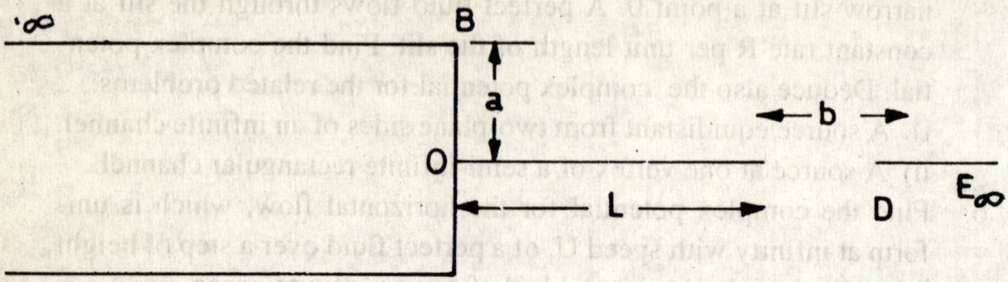

9. Show that the complex potential Ω for a stream, uniform with velocity U at infinity in a horizontal direction $A_\infty B$, flowing over the bed $A_\infty BCD_\infty$ of a semi-infinite ocean is

$$\Omega = \left(\frac{Ua\sqrt{z}}{\pi}\right) \log\left\{\frac{1+w^4}{1-w^4}\right\},$$

$$z = \left(\frac{a}{\pi/2}\right) \left\{\log\left(\frac{1+w}{1-w}\right) + \frac{4w^3}{1-w^4} - 2\tan^{-1} w\right\}$$

where BC = a.

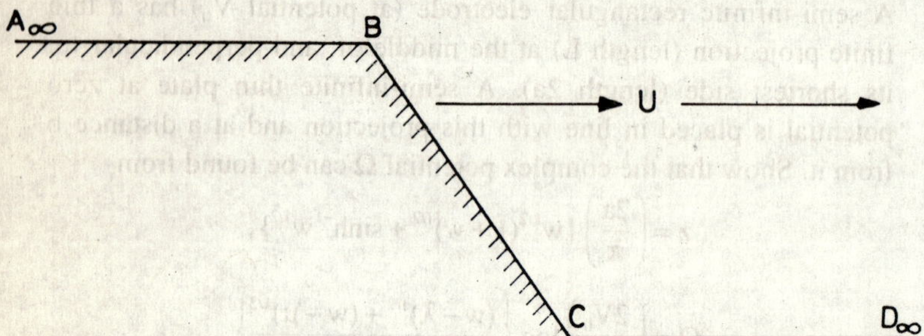

10. A stream which is uniform with velocity U at infinity flows in a parallel-sided channel which has a projection at right angles to one side. Find the complex potential assuming that originally the flow is parallel to the plane sides.

ANSWERS

1. In the half-plane onto which the given region is mapped by the transformation

$w = \sqrt{z^2 - 1} + \cosh^{-1} z$, the temperature is

$$T = \frac{1}{\pi}\tan^{-1}\left(\frac{2xy}{x^2 - y^2 - 1}\right).$$

2. $w = i\pi + \frac{1}{2}\log\frac{z+1}{z-1}.$

3. $w = h(z^2 - 1)^{1/2}.$

4. $w = \frac{ia}{\pi}\cosh^{-1}z.$

5. $\Omega = \frac{R}{\pi}\sinh\left(\frac{\pi w}{2l}\right).$

6. $\Omega = \frac{Uhw}{\pi},\ z = \frac{h}{\pi}\{\sqrt{w^2 - 1} + \cosh^{-1}w\}.$

7. $\Omega = \frac{Uh}{\pi}\log w,\ \frac{dz}{dw} = \left(\frac{h}{\pi w}\right)\sqrt{\frac{w-1}{w-(h/k)^2}}.$

10. $\Omega = \frac{Uh}{\pi}\log w,$

$$z = \frac{h}{\pi}\log\left\{\frac{w(w - \alpha + \sqrt{w^2 - 2\alpha w + 1})}{\alpha w - 1 + \sqrt{w^2 - 2\alpha w + 1}}\right\} + ih.$$

where $\sqrt{\frac{\alpha - 1}{2}} \equiv \tan\left(\frac{k-h}{2h}\pi\right),$

h = width of channel,

k = length of projection.

Chapter 5

Complex Integration

Art 10. RIEMANN'S DEFINITION OF INTEGRATION

Let f(z) be any continuous function of z, not necessarily analytic, defined in a domain D of the complex plane; let C be any curve joining any two points A(z = a) and B(z = b) of D such that C lies entirely within D (Fig. 56), and let $a = z_0, z_1, z_2, \ldots z_{n-1}, z_n = b$ be of C. Then the polygonal line joining these points has length.

$$l_n = |z_1 - z_0| + |z_2 - z_1| + \ldots + |z_n - z_{n-1}|$$

$$= \sum_{k=1}^{n} |z_k - z_{k-1}| \qquad (10.1)$$

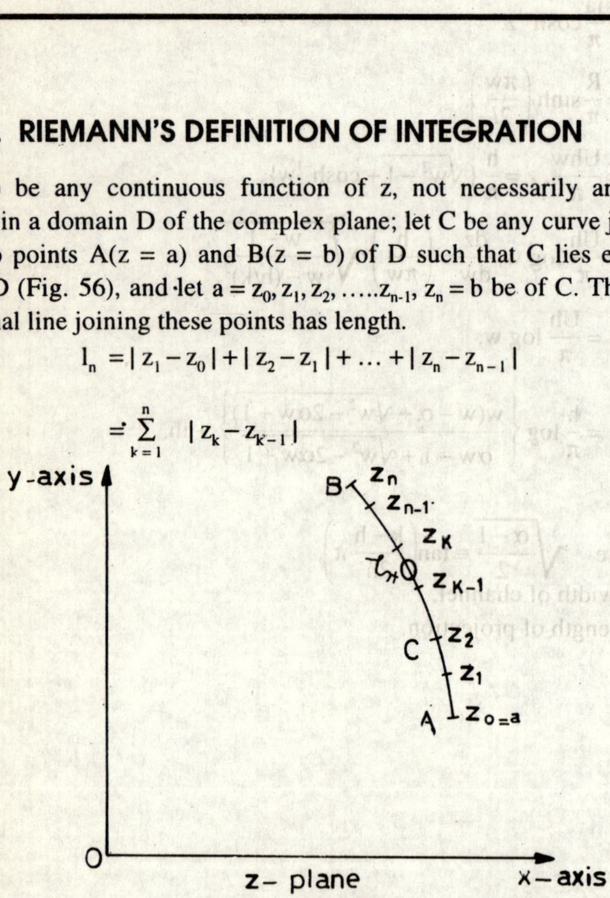

Fig. 56.

If l_n tends to a finite limit l as all the lengths $|z_k - z_{k-1}|$, k = 1, 2,....,n of the intervals tend to zero, *i.e.*, as $n \to \infty$, then C is said to be rectifiable and to have length l. If ζ_k is any point in the subinterval (z_{k-1}, z_k) whose end points are z_{k-1} and z_k where k = 1, 2,....n, and the sum

$$S_n = \sum_{k=1}^{n} (z_k - z_{k-1})\, f(\zeta_k) \tag{10.2}$$

tends to a unique limit S when the number of points of division increases indefinitely in such a manner that the lengths $| z_k - z_{k-1} |$ $k = 1, 2, \ldots, n$ of intervals tend to zero, then S is called the *definite integral* of f(z) along the curve C and is written as

$$S = \int_C f(z) = \lim_{n \to \infty} \sum_{k=1}^{n} (z_k - z_{k-1})\, f(\zeta_k) \tag{10.3}$$

It immediately follows from the definition that if E is any point on C then

$$\int_C f(z)\, dz = \int_{AB} f(z)\, dz = \int_{AE} f(z) dz + \int_{EB} f(z) dz \tag{10.4}$$

where all integrals are taken along the curve C, and also

$$\int_{AE} f(z)\, dz = -\int_{EA} f(z)\, dz. \tag{10.5}$$

Example (i) Evaluate

$$S = \int_{AB} dz.$$

Solution: From the definition of integration, we have

$$S = \lim_{n \to \infty} \sum_{k=1}^{n} (z_k - z_{k-1})$$

since $f(z) = 1$.

or
$$S = \lim_{n \to \infty} \;[(z_1 - z_0) + (z_2 - z_1) + \ldots (z_n - z_{n-1})]$$

$$= \lim_{n \to \infty} (z_n - z_0)$$

$$= b - a \tag{10.6}$$

where $a = z_0$ is the initial point and $b = z_n$ is the final point of curve AB.

Corollary: If the curve AB is closed, *i.e.*, the initial point A = final point B = a(say), the above integral then becomes

$$S = \int_{AB} dz = 0. \tag{10.7}$$

Example (ii) Evaluate

$$S = \int_{AB} z\, dz.$$

Solution: We obtain from the definition

$$S = \lim_{n \to \infty} \sum_{k=1}^{n} (z_k - z_{k-1})\, \zeta_k$$

since $f(z) = z$.

If we take ζ as the point z, then we have

$$S = \lim_{n \to \infty} \sum_{k=1}^{n} (z_k - z_{k-1})\, z_k \tag{10.8}$$

On the other hand, if ζ_k is replaced by the point z_{k-1} then we get

$$S = \lim_{n \to \infty} \sum_{k=1}^{n} (z_k - z_{k-1})\, z_{k-1}. \tag{10.9}$$

Adding (10.8) and (10.9), we obtain

$$2S = \lim_{n \to \infty} \sum_{k=1}^{n} (z_k^2 - z_{k-1}^2).$$

$$= \lim_{n \to \infty} [(z_1^2 - z_0^2) + (z_2^2 - z_1^2) + \ldots + (z_n^2 - z_{n-1}^2)]$$

$$= \lim_{n \to \infty} (z_n^2 - z_0^2)$$

$$= b^2 - a^2.$$

or $$S = \frac{1}{2}(b^2 - a^2). \tag{10.10}$$

In case the curve AB is closed, we have the result

$$S = \int_{AB} z\, dz = 0. \tag{10.11}$$

Example (iii) What is the value of

$$\int_C \frac{dz}{(z-a)^{n+1}}$$

if a is the centre and r is the radius of circle C?

Solution: We have

$$z - a = r\, e^{i\theta} \ (0 \le \theta \le 2\pi)$$

as the equation of given circle C. Thus $dz = ri\, e^{i\theta} d\theta$.

The above integral then takes the form

$$\int_0^{2\pi} \frac{re^{i\theta} d\theta}{r^{n+1} e^{i(n+1)\theta}}$$

$$= \frac{i}{r^n}\int_0^{2\pi} e^{-in\theta} d\theta.$$

If n is a number equal to zero, then it becomes

$$i\int_0^{2\pi} d\theta = 2\pi i. \tag{10.12}$$

On the other hand, if n is a non-zero number, then we have

$$\frac{i}{r^n}\int_0^{2\pi} (\cos n\theta - i \sin n\theta)\, d\theta = 0. \tag{10.13}$$

Contour: A *contour* is a continuous curve consisting of a finite number of arcs, defined by the equations $x = \phi(t)$, $y = \psi(t)$ where the derivatives $\phi'(t), \psi'(t)$ are continuous. It is said to be *closed* if and only if the initial point of first arc and the final point of last arc coincide together. A few examples of closed contours are circles, squares etc.

If C is a closed contour, then the integral of f(z) is written as $\int_C f(z)dz$. It, however, does not indicate the direction in which the contour is described, but we shall always take the counter-clockwise direction, unless otherwise stated.

Properties of Integrals

Likewise not only integrals in real-variables, complex integrals also have the following easily proved properties:

(i) $$\int_C [\phi(z) + \psi(z)]\, dz = \int_C \phi(z)dz + \int_C \psi(z)dz \tag{10.14}$$

(ii) $$\int \phi(z)dz = -\int_{-C} \phi(z)dz, \quad \text{where } -c \text{ indicates} \tag{10.15}$$

the direction opposite to the direction of c.

iii) If k is a constant (real or complex), then

$$\int_C k\phi(z)dz = k\int_C \phi(z)dz. \tag{10.16}$$

(iv) $$\int_C \phi(z)dz = \int_{C_1} \phi(z)dz + \int_{C_2} \phi(z)dz + \int_{C_3} \phi(z)dz, \tag{10.17}$$

where c_1, c_2, c_3 are three different arcs constituting c.

An Important Inequality

If M is the upper bound (maximum value) of $|f(z)|$ on a contour, then from (10.2), we have

$$|S_n| = \left|\sum_{K=1}^{n} (z_K - z_{k-1}) \quad f(\zeta_k)\right|$$

$$\leq \sum_{k=1}^{n} |z_k - z_{k-1}| \quad |f(\zeta_k)|$$

since the modulus of sum of any complex numbers is less than or equal to sum of their moduli. But $|f(\zeta_k)| \leq M$ (follows from the definition of upper bound), the above inequlity becomes

$$|S_n| \leq M \sum_{k=1}^{n} |z_k - z_{k-1}| = Ml_n.$$

Now as $n \to \infty$, we have

$$\left|\int_C f(z)\, dz\right| \leq Ml \tag{10.18}$$

where l is the length of contour C.

It is an important result in the theory of complex integration and will be used several times in the later part of this chapter.

We shall now state and prove a fundamental theorem in the theory of complex analytic functions, known as the *Cauchy-Goursat Theorem.**

Statement: If f(z) is analytic on and inside a closed contour then

$$\int_C f(z)dz = 0. \tag{10.19}$$

Before we proceed with actual proof of this theorem, we shall first establish a Lemma.

Lemma: Given any arbitrarily small positive number ε, it is always possible to divide up the region inside the closed contour into a finite number of squares and partial squares such that within each subregion there exists a point z_0 satisfying the inequality

$$\left|\frac{f(z) - f(z_0)}{z - z_0} - f'(z_0)\right| < \varepsilon \tag{10.20}$$

for all points z belonging to subregion.

To prove it, we divide the region inside C into equal squares each of side d by drawing lines parallel to real and imaginary axes at a constant distance d (Fig. 57). We, thus, obtain squares which lie completely

*It dates from the year 1825 and forms the basis of the whole theory of functions of a complex variable.

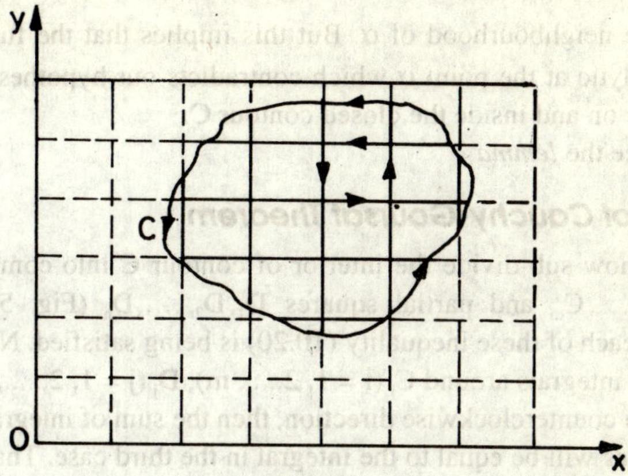

Fig. 57

within C and some squares lying partially inside C. Thus we have divided the whole region inside C into smaller regions, Of the sub-regions so obtained some may satisfy the inequality (10.20), while others do not. Leaving aside the sub-regions satisfying (10.20), we sub-divide the others, not satisfying (10.20), into smaller sub-regions by lines joining the mid-points of their opposite sides. If there are still some more sub-regions which do not satisfy (10.20), we further sub-divide them in the same manner.

Thus we ultimately arrive at either of the following points:

i) There is no sub-region left which do not satisfy (10.20) after a finite number of steps of operation

or

ii) the process continues indefinitely.

In the later case, there always exists a sub-regions which do not satisfy (10.20). In this case, we obtain an unending sequence of sub-regions each included in the preceding one and for each of which (10.20) is not being satisfied. This sequence will certainly have some limiting point (say) α lying within C.

As at this point, (10.20) is not satisfied, we have

$$\left| \frac{f(z) - f(z_0)}{z - z_0} - f'(z_0) \right| \nless \varepsilon$$

in some neighbourhood of α. But this implies that the function f(z) is not analytic at the point α which contradicts our hypothesis that f(z) is analytic on and inside the closed contour C.

Hence the *lemma*.

Proof of Cauchy-Goursat Theorem

We now sub-divide the interior of contour C into complete squares $C_1, C_2, \ldots, C_m$ and partial squares $D_1, D_2, \ldots, D_n$ (Fig. 56) such that within each of these inequality (10.20) is being satisfied. Now assuming that the integrals around C_i (i = 1, 2,...., m), D_j (j = 1, 2,....., n) and C are taken in counterclockwise direction, then the sum of integrals in the first two cases will be equal to the integral in the third case. That is

$$\sum_{i=1}^{m} \int_{C_i} f(z)dz + \sum_{j=1}^{n} \int_{D_j} f(z)dz = \int_{C} f(z)dz. \tag{10.21}$$

From the lemma, we have

$$\left| \frac{f(z) - f(z_0)}{z - z_0} - f'(z_0) \right| < \varepsilon,$$

or $$f(z) = f(z_0) + (z - z_0) f'(z_0) + \beta(z - z_0) \tag{10.22}$$

where $|\beta| < \varepsilon$.

Now

$$\int_{C_i} f(z)\,dz = \int_{C_i} [\,f(z_0) + (z - z_0) f'(z_0) + \beta(z - z_0)]\,dz$$

$$= \{f(z_0) - z_0 f'(z_0)\} \int_{C_i} dz + f'(z_0) \int_{C_i} z dz$$

$$+ \int_{C_i} (z - z_0)\,\beta\,dz.$$

Employing the results of examples (i), (ii) for closed contours, it gives

$$\int_{C_i} f(z)\,dz = \int_{C_i} (z - z_0)\,\beta\,dz. \tag{10.23}$$

Proceeding along the same lines, we obtain for D_j

$$\int_{D_j} f(z)\,dz = \int_{D_j} (z - z_0)\,\beta\,dz. \tag{10.24}$$

Putting these values of $\int_{C_i} f(z)dz$ and $\int_{D_j} f(z)dz$ in (10.21), we get

$$\int_C f(z)\, dz = \sum_{i=1}^{m} \int_{C_i} (z - z_0)\, \beta\, dz + \sum_{j=1}^{n} \int_{D_j} (z - z_0)\, \beta dz$$

whence

$$\left| \int_C f(z)dz \right| \le \sum_{i=1}^{m} \left| \int_{C_i} (z - z_0)\beta dz \right| + \sum_{j=1}^{n} \left| \int_{D_j} (z - z_0)\, \beta dz \right|. \qquad (10.25)$$

Now

$$\left| \int_{C_i} (z - z_0)\beta dz \right| \le \int_{C_i} | z - z_0 |\ \ | \beta |\ \ | dz | < l_i\sqrt{2}\,.\,\varepsilon \int_{C_i} | dz |$$

$$< \varepsilon . l_i \sqrt{2}\,.\ \ 4l_i = 4\sqrt{2}\ \ A_i. \qquad (10.26)$$

Here we have made use of the facts that $| z - z_0 | \le l_i\sqrt{2}$, the length of diagonal of i-th square; $| \beta | < \varepsilon$; $\int_{C_i} | dz | = 4\, l_i$ and $A_i = l_i .\, l_i$, the area of i-th square.

Similarly, if l'_j denotes the length of side and A'_j, the area of square of which D_j is a part, we have

$$\left| \int_{D_j} (z - z_0)\beta dz \right| \le \int_{D_j} | z - z_0 |\ \ | \beta |\ \ | dz |$$

$$< \varepsilon\, l'_j \sqrt{2} \int_{D_j} | dz | .$$

But $\int_{D_j} | dz |$ is equal to the perimeter of D_j which is less than $(4\, l_j + s_j)$, s_j being that part of which belongs to D_j. Thus

$$\left| \int_{D_j} f(z)dz \right| < \varepsilon\sqrt{2}\, l'_j\, (4\, l'_j + s_j)$$

$$< \varepsilon\sqrt{2}\, (4\, A'_j + l'_j\, s_j)$$

$$< \varepsilon\sqrt{2}\, (4\, A'_j + d\, s_j), \qquad (10.27)$$

since $l'_j \le d$, d being the side of the squares with which the region was originally covered.

Now employing (10.26), (10.27) in (10.25), we obtain

$$\left|\int_C f(z)dz\right| < 4\sqrt{2}\varepsilon \sum_{i=1}^{m} A_i + \varepsilon\sqrt{2}\sum_{j=1}^{n}(4A'_j + ds_j)$$

$$< 4\sqrt{2}\varepsilon\left\{\sum_{i=1}^{m} A_i + \sum_{j=1}^{n} A'_j\right\} + \varepsilon\sqrt{2}.d\sum_{j=1}^{n} s_j$$

$$< \varepsilon\sqrt{2}\,(4A + dl).$$

$$< k\varepsilon$$

since $\sum_{i=1}^{n} s_j = l$, the length of contour C.

Here A denotes the combined area of the squares of side d with which the region was originally covered and k is any constant.

But ε, being an arbitrary small positive number, can be chosen as small as we desire, hence

$$\int_C f(z)dz = 0$$

which completes the proof of this *theorem*.

Generalisation of Cauchy-Goursat Theorem

Statement: If C is a simple closed contour enclosing a second such contour γ, and if the function f(z) is analytic in the region between them, then

$$\int_C f(z)dz = \int_\gamma f(z)dz \tag{10.28}$$

where both integrals are taken in counter-clockwise direction.

Proof: To prove it, we consider a curve AB joining the contours C and γ and lying between them (Fig. 58). Now if Γ is the curve C + AB + γ + BA indicated by the arrows, then by Cauchy's theorem we have

$$\int_\gamma f(z)\,dz = 0.$$

But

$$\int_\gamma f(z)dz = \int_C f(z)dz + \int_{AB} f(z)dz - \int_\gamma f(z)dz + \int_{BA} f(z)dz$$

$$= \int_C f(z)dz - \int_\gamma f(z)dz$$

by taking integral over γ in counter-clockwise sense and using (10.5);

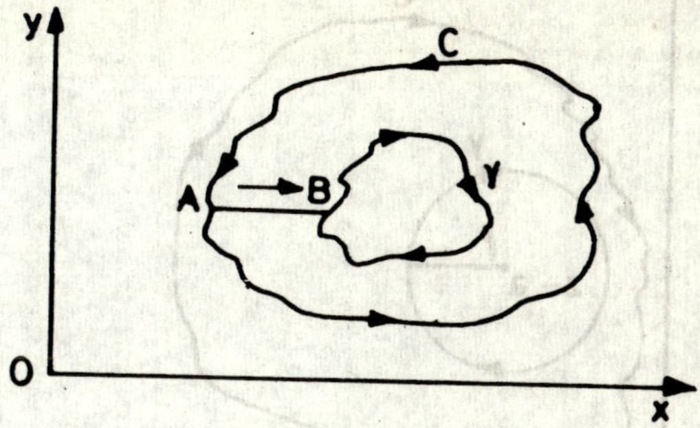

Fig. 58

whence

$$\int_C f(z)dz = \int_\gamma f(z)dz.$$

An immediate generalization of this result for n-closed contours $\gamma_1, \gamma_2, \ldots, \gamma_n$ lying within C is

$$\int_C f(z)dz = \sum_{i=1}^{n} \int_{\gamma_i} f(z)dz. \tag{10.29}$$

Cauchy's Integral Formula

Statement: If f(z) is analytic on and inside a closed contour C, and if z = a is any point within C, then

$$f(a) = \frac{1}{2\pi i} \int_C \frac{f(z)}{z-a} dz. \tag{10.30}$$

Proof: This theorem expresses the fact that if a function is analytic within and on a closed contour then its value at every interior point can be evaluated.

With the point z = a as centre and P as the radius, we construct a small circle γ lying entirely within C (Fig. 59(a)).

The function $\frac{f(z)}{z-a}$ is then analytic between and on the contours C and γ. Thus

by generalisation of Cauchy-Goursat theorem (10.28), we have

$$\int_C \frac{f(z)}{z-a} dz = \int_\gamma \frac{f(z)}{z-a} dz$$

$$= \int_\gamma \frac{f(z) - f(a)}{z-a} dz + f(a) \int_\gamma \frac{dz}{z-a} \tag{10.31}$$

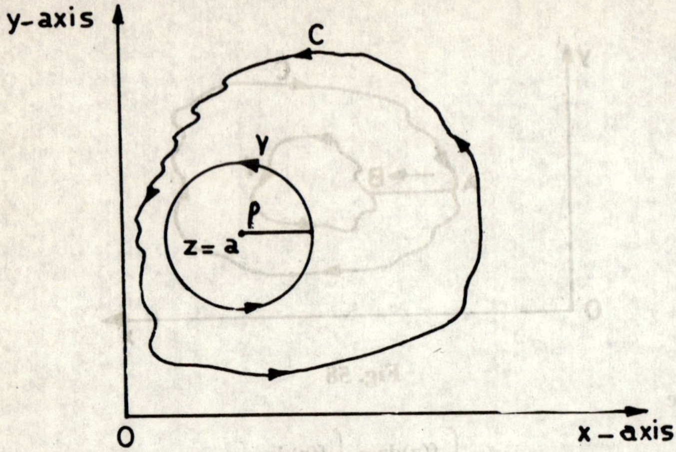

Fig. 59(a)

Employing the result of example (iii) for n = 0 , *i.e.* (10.12) in this equation, we obtain

$$\int_C \frac{f(z)}{z-a}\,dz = \int_\gamma \frac{f(z)-f(a)}{z-a}\,dz + 2\pi i f(a). \tag{10.32}$$

Since f(z) is analytic at z = a and so it is also continuous at this point, thus for $\varepsilon > 0$

$$|f(z)-f(a)| < \varepsilon$$

for all points z belonging to $\rho = |z-a| < \delta$, δ being a positive number depending on ε and ρ. By using the inequality (10.18), we get

$$\left|\int_\gamma \frac{f(z)-f(a)}{z-a}\,dz\right| \le \int_\gamma \frac{|f(z)-f(a)|}{|z-a|}\,|dz|$$

$$< \frac{\varepsilon}{\rho}\cdot 2\pi\rho = 2\pi\varepsilon, \tag{10.33}$$

so that left hand side of (10.33) tends to the limit zero as ρ approaches zero. Hence (10.32) gives

$$f(a) = \frac{1}{2\pi i}\int_C \frac{f(z)}{z-a}\,dz,$$

which is the required formula.

Derivatives of an Analytic Function

Theorem: If a function f(z) is analytic on and inside a closed contour C, then it possesses a unique first-order derivative

$$f'(a) = \frac{1}{2\pi i} \int_C \frac{f(z)}{(z-a)^2} dz \qquad (10.34)$$

at all points z = a inside C.

Proof: We choose the increment Δa in such a manner that z = a + Δa is also an interior point of C (Fig. 59 (b)). Then by Cauchy's integral formula (10.30), we have

$$f(a) = \frac{1}{2\pi i} \int_C \frac{f(z)}{z-a} dz,$$

$$f(a+\Delta a) = \frac{1}{2\pi i} \int_C \frac{f(z)dz}{z-a-\Delta a},$$

whence $f(a+\Delta a) - f(a) = \dfrac{\Delta a}{2\pi i} \displaystyle\int_C \frac{f(z)dz}{(z-a)(z-a-\Delta a)}$

or

$$\frac{f(a+\Delta a)-f(a)}{\Delta a} = \frac{1}{2\pi i} \int_C \frac{f(z)dz}{(z-a)^2} + \frac{\Delta a}{2\pi i} \int_C \frac{f(z)dz}{(z-a)^2(z-a-\Delta a)}. \qquad (10.35)$$

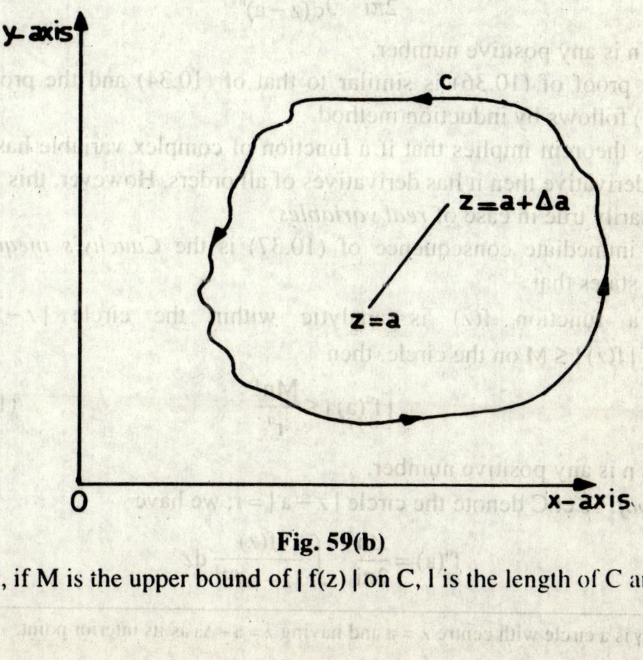

Fig. 59(b)

Now, if M is the upper bound of | f(z) | on C, l is the length of C and d is the shortest distance between any point in the circle γ^* and any point in contour C, then

$$|z-a|>r, \; |z-a-\Delta a|>r.$$

Thus by using (10.18)

$$\left|\frac{\Delta a}{2\pi i}\int_C \frac{f(z)\,dz}{(z-a)^2(z-a-\Delta a)}\right| \le \frac{|\Delta a|}{2\pi}\int_C \frac{|f(z)|\,|dz|}{|z-a|^2|z-a-\Delta a|} < \frac{|\Delta a|}{2\pi}\cdot\frac{Ml}{r^3}$$

which tends to zero as Δa approaches the zero value. Hence (10.35) gives

$$f'(a) = \frac{1}{2\pi i}\int_C \frac{f(z)}{(z-a)^2}\,dz.$$

Corollary: Now prove that

$$f''(a) = \frac{\angle 2}{2\pi i}\int_C \frac{f(z)}{(z-a)^3}\,dz \tag{10.36}$$

and hence in general

$$f^n(a) = \frac{\angle n}{2\pi i}\int_C \frac{f(z)}{(z-a)^{n+1}}\,dz, \tag{10.37}$$

where n is any positive number.

The proof of (10.36) is similar to that of (10.34) and the proof of (10.37) follows by induction method.

This theorem implies that if a function of complex variable has first order derivative then it has derivatives of all orders. However, this is not necessarily true in case of *real variables*.

An immediate consequence of (10.37) is the *Cauchy's inequality* which states that:

If a function f(z) is analytic within the circle $|z-a|=r$ and if $|f(z)| \le M$ on the circle, then

$$|f^n(a)| \le \frac{Mn!}{r^n} \tag{10.38}$$

where n is any positive number.

Proof: Let C denote the circle $|z-a|=r$; we have

$$f^n(a) = \frac{\angle n}{2\pi i}\int_C \frac{f(z)}{(z-a)^{n+1}}\,dz$$

*Here γ is a circle with centre $z = a$ and having $z = a + \Delta a$ as its interior point.

whence

$$|f^n(a)| = \left|\frac{\angle n}{2\pi i}\int_C \frac{f(z)}{(z-a)^{n+1}}\,dz\right|$$

$$\leq \frac{\angle n}{2\pi}\int_C \frac{|f(z)|}{|z-a|^{n+1}}\,|dz|$$

Employing $|f(z)| \leq M$ on C and the equation of circle $z-a=r\,e^{i\theta}$ $(0 \leq \theta \leq 2\pi)$, it gives

$$|f^n(a)| \leq \frac{M\angle n}{2\pi}\int_0^{2\pi}\frac{r\,d\theta}{r^{n+1}}$$

$$= \frac{M\angle n}{2\pi r^n}\,.\,2\pi$$

hence

$$|f^n(a)| \leq \frac{M\angle n}{r^n}.$$

In other words, Cauchy's inequality states that a function cannot have a maximum value at any point where it is analytic.

Poisson's Integral Formula

If f(z) is analytic in $|z| < \rho$, then

$$f(re^{i\theta}) = \frac{1}{2\pi}\int_0^{2\pi}\frac{(R^2-r^2)f(Re^{i\theta})}{R^2-2Rr\cos(\theta-\phi)+r^2}\,d\phi \qquad (10.39)$$

where $r\,e^{i\theta}$ is any point of the domain $|z| < \rho$ and R is any number such that $r < R < \rho$.

Proof: Supposing that $z = r\,e^{i\theta}$ is any point of the domain $|z| < \rho$ where $|z| = R$ is any circle C lying in this domain such that $r < R < \rho$. (Fig. 60).

By Cauchy's integral formula (10.30), we have

$$f(z) = \frac{1}{2\pi i}\int_C \frac{f(\zeta)}{\zeta - z}\,d\zeta. \qquad (10.40)$$

The inverse point of z with respect to circle C is $\frac{R^2}{\bar{z}}$ and it lies outside this

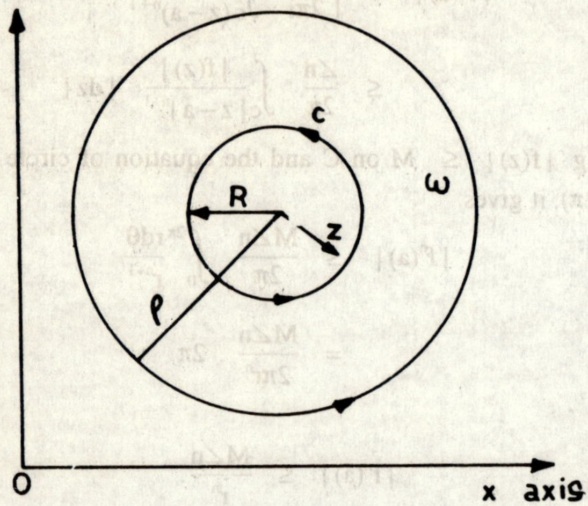

Fig. 60

circle. Thus, the function $\dfrac{f(\zeta)}{\zeta - \frac{R^2}{\bar{z}}}$ is analytic within C and hence by Cauchy-Goursat theorem (10.19), we have

$$0 = \int_C \frac{f(\zeta)}{\zeta - \frac{R^2}{\bar{z}}}\, d\zeta$$

or
$$0 = \frac{1}{2\pi i} \int_C \frac{f(\zeta)}{\zeta - \frac{R^2}{\bar{z}}}\, d\zeta. \qquad (10.41)$$

Subtraction of (10.41) from (10.40) gives

$$f(z) = \frac{1}{2\pi i} \int_C \frac{(R^2 - z\bar{z}) f(\zeta)}{(\zeta - z)(R^2 - \bar{z}\zeta)}\, d\zeta.$$

If we put $z = re^{i\theta}, \zeta = Re^{i\phi}$, we get

$$f(re^{i\theta}) = \frac{1}{2\pi i} \int_0^{2\pi} \frac{(R^2 - r^2) f(Re^{i\phi}) Re^{i\phi} i\, d\phi}{(Re^{i\phi} - re^{i\theta})(R^2 - Re^{i\phi} re^{-i\theta})}. \qquad (10.42)$$

It simplifies to

$$f(re^{i\theta}) = \frac{1}{2\pi} \int_0^{2\pi} \frac{(R^2 - r^2)f(Re^{i\phi})}{R^2 - 2Rr\cos(\theta - \phi) + r^2} d\phi$$

which is known as *Poisson's Integral Formula.*

Finally, if $f(re^{i\theta})$ is replaced by

$u(r,\theta) + iv(r,\theta)$ and $f(Re^{i\phi})$ by $u(R,\phi) + iv(R,\phi)$
then by equating real and imaginary parts, we obtain

$$u(r,\theta) = \frac{1}{2\pi} \int_0^{2\pi} \frac{(R^2 - r^2)u(R,\phi)}{R^2 - 2Rr\cos(\theta - \phi) + r^2} d\phi \qquad (10.43)$$

$$v(r,\theta) = \frac{1}{2\pi} \int_0^{2\pi} \frac{(R^2 - r^2)v(R,\phi)}{R^2 - 2Rr\cos(\theta - \phi) + r^2} d\phi. \qquad (10.44)$$

The formulae (10.43) an (10.44) express the real and imaginary parts of the value of function at any interior point in terms of the real and imaginary parts of the values of functions at points on the bounding circles.

Definition: A function, analytic at all points of the region where it is defined, is called an *integral function.* We shall now state and prove a theorem based on such type of functions and known as *Liouville's theorem*. It states that if an integral function is bounded for all values of z then it is a constant.

Proof: Let z_1, z_2 be two distinct but arbitrary points of z-plane where the integral function f(z) is defined. With origin as centre and radius r, we draw a circle C such z_1, z_2 are its interior points. Thus we have $|z_1| < r$ and $|z_2| < r$. Applying Cauchy's integral formula (10.30) at these points, we obtain

$$f(z_1) = \frac{1}{2\pi i} \int_C \frac{f(z)}{z - z_1} dz,$$

$$f(z_2) = \frac{1}{2\pi i} \int_C \frac{f(z)}{z - z_2} dz.$$

Hence

$$|f(z_1) - f(z_2)| = \left| \frac{1}{2\pi i} \int_C \frac{(z_1 - z_2)f(z)}{(z - z_1)(z - z_2)} dz \right|$$

$$\leq \frac{1}{2\pi} \int_C \frac{|z_1 - z_2| \; |f(z)|}{|z - z_1| \; |z - z_2|} |dz|$$

$$< \frac{1}{2\pi} \frac{|z_1 - z_2| M \quad 2\pi r}{(r - |z_1|)(r - |z_2|)}$$

by using theorem (3) M being the maximum value of $|f(z)|$. The right hand side of this inequality tends to the limit zero as r approaches the infinite values. Hence $f(z_1) = f(z_2)$.

But z_1 and z_2 are arbitrary points of z-plane; therefore f(z) is constant at all points. Hence the *theorem.*

The converse of Cauchy-Goursat theorem is *Morera's theorem* which states:

If a function f(z) is continuous in a region D and if

$$\int_C f(z)dz = 0 \tag{10.45}$$

where C is any closed contour lying inside this region, then f(z) is analytic throughout the interior of D.

Proof: Let a be a fixed point and z any variable point of D and let C_1 and C_2 be any two continuous rectifiable curves in D joining a to z (Fig. 61). Then the curve C constituted by C_1 and C_2 is a closed rectifiable continuous curve in D.

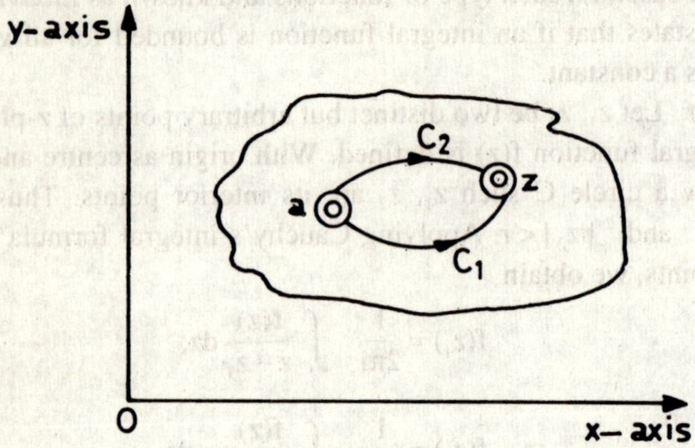

Fig. 61

So by using (10.45), we get

$$\int_{C_1 - C_2} f(z)\,dz = \int_{C_1} f(z)\,dz - \int_{C_2} f(z)dz$$

$$= 0$$

or $\int_{C_1} f(z)dz = \int_{C_2} f(z)dz$

which implies that the integral along any curve in D joining the points a and z is the same.

Thus we may write

$$F(z) = \int_a^z f(\zeta)\, d(\zeta), \tag{10.46}$$

where ζ is the variable of integration.

Now

$$\frac{F(z+h) - F(z)}{h} - f(z)$$

$$= \frac{1}{h}\left[\int_a^{z+h} f(\zeta)\, d\zeta - \int_a^z f(\zeta) d\zeta - f(z)\int_z^{z+h} d\zeta\right|$$

$$= \frac{1}{h}\int_z^{z+h} [f(\zeta) - f(z)]\, d\zeta. \tag{10.47}$$

Since $f(\zeta)$ is continuous at z thus for given $\varepsilon > 0$, there exists a positive number δ such that

$$| f(\zeta) - f(z) | < \varepsilon \quad \text{for} \quad |\zeta - z| < \delta. \tag{10.48}$$

Hence $\left|\frac{F(z+h) - F(z)}{h} - f(z)\right|$

$$= \left|\frac{1}{h}\int [f(\zeta) - f(z)]\, d\zeta\right|$$

$$\leq \frac{1}{|h|}.\, \varepsilon.\, |h| = \varepsilon$$

for $|h| < \delta$.

Thus

$$\lim_{h \to 0} \frac{F(z+h) - F(z)}{h} = F'(z)$$

i.e., $F'(z) = f(z)$.

This relation holds good for all values of z

Thus we observe that f(z) appears as the derivative of a function and hence f(z) is analytic.

This proves the theorem.

Development of Analytic Functions as Power Series

Taylor's Theorem

If a function f(z) is analytic on and inside a circle C with its centre at z = a and radius r and if ζ is any point interior to C, then

$$f(\zeta) = \sum_{n=0}^{\infty} f^n(a) \frac{(\zeta - a)^n}{\angle n} \tag{10.49}$$

where $f^0(a) = f(a)$, $\angle 0 = 1$.

Proof: By Cauchy's integral formula (10.30), we have

$$f(\zeta) = \frac{1}{2\pi i} \int_C \frac{f(z)}{z - \zeta} \, d\zeta. \tag{10.50}$$

Now let

$$S = \frac{1}{z-a} \left[\frac{1 - (\zeta - a)^n/(z - a)^n}{1 - (\zeta - a)/(z - a)} \right]$$

$$= \frac{1}{z-\zeta} - \frac{(\zeta - a)^n}{(z - a)^n (z - \zeta)}.$$

Hence

$$\frac{1}{z-\zeta} = S + \frac{(\zeta - a)^n}{(z - a)^n (z - \zeta)}. \tag{10.51}$$

Employing this value in (10.50), we obtain

$$f(\zeta) = \frac{1}{2\pi i} \int_C \left[\frac{1}{(z-a)} + \frac{\zeta - a}{(z-a)^2} + \ldots + \frac{(\zeta - a)^{n-1}}{(z-a)^{n-1}} + \frac{(\zeta - a)^n}{(z - a)^n (z - \zeta)} \right] f(z) dz$$

$$= f(a) + \frac{(\zeta - a)}{\angle 1} f'(a) + \frac{(\zeta - a)^2}{\angle 2} f''(a) + \ldots + \frac{(\zeta - a)^{n-1}}{\angle n-1} f^{n-1}(a) + R_n \tag{10.52}$$

on using (10.37) and putting

$$R_n = \frac{1}{2\pi i}\int_C \frac{(\zeta - a)^n f(z)}{(z-a)^n (z-\zeta)} dz.$$

Now $|z-\zeta| = |z-a-\zeta+a| \geq |z-a| - |\zeta - a| = r - |\zeta - a|$,

and $|f(z)| \leq M$ on C where M is the maximum value of $|f(z)|$ on C

Thus we have

$$|R_n| \leq \frac{1}{2\pi}\int_C \left|\frac{\zeta - a}{z-a}\right|^n \frac{|f(z)|}{|z-\zeta|}|dz|$$

$$\leq \frac{1}{2\pi} . k^n \frac{M}{r - |\zeta - a|} . 2\pi r$$

$$= \frac{Mr}{r - |\zeta - a|} k^n$$

where $\left|\frac{\zeta - a}{z - a}\right| = k < 1$ for any point z on C. Hence, the limit of $|R_n|$ tends to

zero as n approaches infinite values. Thus (10.52) gives

$$f(\zeta) = f(a) + \frac{(\zeta - a)}{\angle 1} f'(a) + \frac{(\zeta - a)^2}{\angle 2} f''(a)$$

$$+ \ldots + \frac{(\zeta - a)^n}{\angle 2} f^n(a) + \ldots \quad (10.53)$$

which completes the proof of this theorem.

A Particular Case

If we put a = 0 in (10.53), we get

$$f(\zeta) = f(0) + \frac{\zeta}{\angle 1} f'(0) + \frac{\zeta^2}{\angle 2} f''(0) + \ldots$$

$$+ \frac{\zeta^n}{\angle n} f^n(0) + \ldots \quad (10.54)$$

This is called *Maclaurin Series*.

Example. Prove that

$$\sin z = z - \frac{z^3}{3} + \ldots$$

Solution: Since f(z) = sin z is analytic in any region, we may develop it about the origin. Hence by Maclaurin series (10.54), we get

$$f(z) = f(0) + zf'(0) + \frac{z^2}{\angle 2} f''(0) + \ldots$$

But $f(0) = 0, f'(0) = 1, f''(0) = 0$ and so on by ordinary differentiation whence we obtain

$$\sin z = z - \frac{z^3}{\angle 3} + \ldots$$

There is a more general type of expansion valid for a function analytic in an annulus. It is known as the *Laurent expansion*. Its statement is

Let C_1, C_2 be two concenteric circles with common centre at $z = a$ and radii r_1, r_2 such that $r_1 > r_2$, respectively, and let D be the annular domain between these circles. If $f(z)$ is analytic on C_1 and C_2 and within D and if $z = \zeta$, is a point of D, then

$$f(\zeta) = \sum_{k=0}^{\infty} a_k(\zeta - a)^k + \sum_{k=1}^{\infty} b_k(\zeta - a)^{-k} \tag{10.55}$$

where

$$a_k = \frac{1}{2\pi i} \int_{C_1} \frac{f(z)}{(z-a)^{k+1}}\, dz, \tag{10.56}$$

$$b_k = \frac{1}{2\pi i} \int_{C_2} \frac{f(z)}{(z-a)^{1-k}}\, dz. \tag{10.57}$$

Proof: Let γ be a circle with centre $z = \zeta$ and radius ρ such that all points of γ are interior points of D (Fig. 62). If D_1 is the domain consisting of D excluding the interior of γ, then within D_1 and on its boundaries, *i.e.*, C_1, C_2, γ, the function $\dfrac{f(z)}{z-\zeta}$ is analytic. Whence by generalisation of Cauchy-Goursat theorem (10.28), we get

$$\int_{C_1} \frac{f(z)}{z-\zeta}\, dz = \int_{C_2} \frac{f(z)}{z-\zeta}\, dz + \int_{\gamma} \frac{f(z)}{z-\zeta}\, dz$$

where the integrals are taken in counter-clockwise sense. But by Cauchy's integral formula (10.30)

$$f(\zeta) = \frac{1}{2\pi i} \int_{\gamma} \frac{f(z)}{z-\zeta}\, dz,$$

thus

$$f(\zeta)=\frac{1}{2\pi i}\int_{C_1}\frac{f(z)}{z-\zeta}dz-\frac{1}{2\pi i}\int_{C_2}\frac{f(z)}{z-\zeta}dz. \qquad (10.58)$$

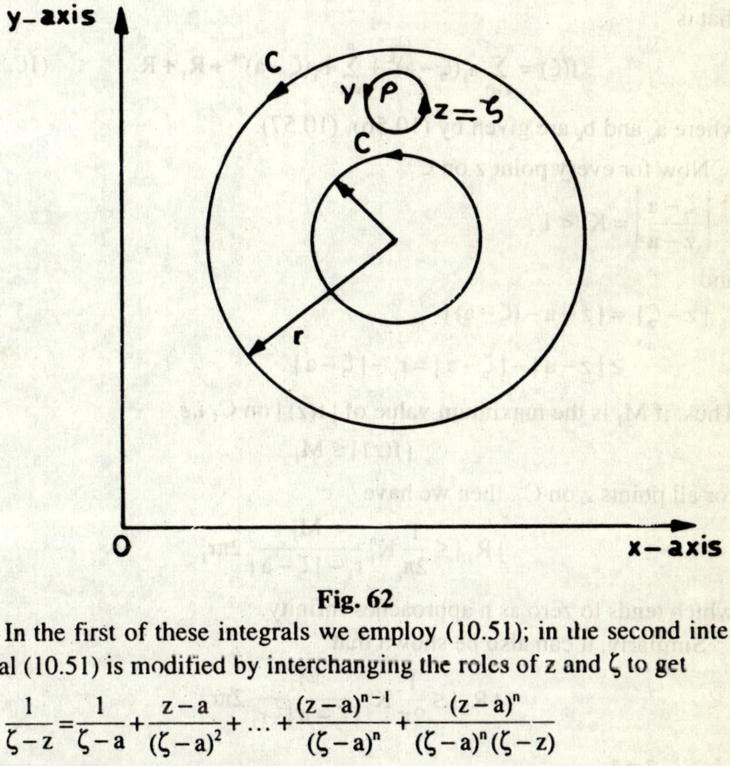

Fig. 62

In the first of these integrals we employ (10.51); in the second integral (10.51) is modified by interchanging the roles of z and ζ to get

$$\frac{1}{\zeta-z}=\frac{1}{\zeta-a}+\frac{z-a}{(\zeta-a)^2}+\ldots+\frac{(z-a)^{n-1}}{(\zeta-a)^n}+\frac{(z-a)^n}{(\zeta-a)^n(\zeta-z)}$$

whence (10.58) takes the form

$$f(\zeta)=\frac{1}{2\pi i}\int_{C_1}\left[\sum_{k=0}^{n-1}\frac{(\zeta-a)^k}{(z-a)^{k+1}}\right]f(z)dz+R_1$$

$$+\frac{1}{2\pi i}\int_{C_2}\left[\sum_{k=1}^{n}\frac{(z-a)^{k-1}}{(\zeta-a)^k}\right]f(z)dz+R_2$$

where

$$R_1 = \frac{1}{2\pi i} \int_{C_1} \frac{(\zeta - a)^n f(z) dz}{(z-a)^n (z-\zeta)},$$

$$R_2 = \frac{1}{2\pi i} \int_{C_2} \frac{(z-a)^n f(z) dz}{(\zeta - a)^n (\zeta - z)}.$$

that is

$$f(\zeta) = \sum_{k=0}^{n-1} a_k (\zeta - a)^k + \sum_{k=1}^{n} b_k (\zeta - a)^{-k} + R_1 + R_2 \qquad (10.59)$$

where a_k and b_k are given by (10.56), (10.57).

Now for every point z on C_1,

$$\left| \frac{\zeta - a}{z - a} \right| = K_1 < 1$$

and

$$|z - \zeta| = |z - a - (\zeta - a)|$$

$$\geq |z - a| - |\zeta - a| = r_1 - |\zeta - a|.$$

Thus, if M_1 is the maximum value of $|f(z)|$ on C_1 *i.e.*

$$|f(z)| \leq M_1$$

for all points z on C_1, then we have

$$|R_1| \leq \frac{1}{2\pi} K_1^n \frac{M_1}{r_1 - |\zeta - a|} 2\pi r_1$$

which tends to zero as n approaches infinity.

Similarly, it can also be shown that

$$|R_2| \leq \frac{1}{2\pi} K_2^n \frac{M_2}{|\zeta - a| - r_2} 2\pi r_2$$

where $\left|\frac{z-a}{\zeta - a}\right| = k_2$ and $|f(z)| \leq M_2$ for all z on C_2. Trivially the limit of $|R_2|$ tends to $|R_2|$ tends to zero as n approaches infinite values. Hence, (10.59) reduces to (10.55) which completes the proof of the theorem.

A Particular Case

If f(z) is also analytic in C_2, then Laurent expansion (10.55) reduces to the Taylor's form (10.49) where $a_k = \dfrac{f^k(a)}{k}$.

Example. Prove that $\sin\left\{c\left(z+\frac{1}{z}\right)\right\}$ can be expanded in a series of the form

$$\sum_{k=0}^{\infty} a_k z^k + \sum_{k=1}^{\infty} b_k z^{-k}$$

where

$$a_k = b_k = \frac{1}{2\pi}\int_0^{2\pi} \sin(2c\cos\theta)\cos k\theta\, d\theta.$$

Solution: The only point at which the function $\left(z+\frac{1}{z}\right)$ is not analytic is the origin. Thus $\sin\left\{c\left(z+\frac{1}{z}\right)\right\}$ is analytic in the annulus $r_2 \le |z| \le r_1$ irrespective of the largeness of r_1 and smallness of r_2. Hence by Laurent's expansion (10.55) to (10.57)

$$\sin\left\{c\left(z+\frac{1}{z}\right)\right\} = \sum_{k=0}^{\infty} a_k z^k + \sum_{k=1}^{\infty} b_k z^{-k}$$

where

$$a_k = \frac{1}{2\pi i}\int_{C_1} \sin\left\{c\left(z+\frac{1}{z}\right)\right\}\frac{dz}{z^{k+1}},$$

$$b_k = \frac{1}{2\pi i}\int_{C_2} \sin\left\{c\left(z+\frac{1}{z}\right)\right\}\frac{dz}{z^{1-k}},$$

C_1, C_2 being any circles with origin as the common centre.

Let C_1 be the unit circle $|z| = 1$. Thus $z = e^{i\theta} (0 \le \theta \le 2\pi)$ is the equation of circle whence

$$a_k = \frac{1}{2\pi i}\int_0^{2\pi i} \sin(2c\cos\theta)\, e^{-ni\theta} d\theta$$

$$= \frac{1}{2\pi i}\int_0^{2\pi} \sin(2c\cos\theta)\cos n\theta d\theta$$

because $\int_0^{2\pi} \sin(2c\cos\theta)\sin n\theta\, d\theta = 0.$

But if z is replaced by $\frac{1}{z}$, the function $\sin\left\{c\left(z+\frac{1}{z}\right)\right\}$ remains unaltered, hence

$$b_k = a_k = \frac{1}{2\pi}\int_0^{2\pi} \sin(2c\cos\theta)\cos n\theta\, d\theta$$

which completes the proof.

EXERCISE

1. Evaluate by the definition of integration

$$I = \int_{AB} |dz|$$

where AB is any arc.

2. Evaluate $\int_0^{1+i}(x^2+iy)\,dz$ along the paths $y = x$ and $y = x^2$.

3. What are the values of $\int_C \frac{e^z}{z^2+1}dz$ if C is a unit circle with centre at (i) $z = i$ and (ii) $z = -i$?

4. What are the values of $\int_C \frac{z+1}{z^2(z-2)}dz$ around the circles C?

(i) $|z| = 1$, (ii) $|z-2-i| = 2$, (iii) $|z-1-2i| = 2$?

5. Obtain expressions for. $\frac{1}{(z^2+1)(z^2+2)}$ which are valid when

(i) $|z| < 1$, (ii) $1 < |z| < \sqrt{2}$, (iii) $|z| > \sqrt{2}$.

6. Prove that when $z \neq 0$

$$\cosh\left\{c\left(z+\frac{1}{z}\right)\right\} = \sum_{k=-\infty}^{\infty} a_k z^k$$

where

$$a_k = \frac{1}{2\pi}\int_0^{2\pi} \cosh(2c\cos\theta)\cos k\theta\, d\theta.$$

7. Show that for $z \neq 0$

$$\left[\frac{1}{2}c\left(z-\frac{1}{z}\right)\right] = \sum_{k=-\infty}^{\infty} a_k z^k$$

where $a_k = \frac{1}{2\pi}\int_0^{2\pi} \cos(k\theta - c\sin\theta)\, d\theta.$

8. Evaluate $\int_C 24z^2\,dz$ where C is the line segment form $z = 1+i$ to $z = 3-4i$.

counter-clockwise.

10. Find the value of $\oint_C [ze^2/(4z+\pi i)^2]\,dz$ where C is the unit circle traversed in the counter clockwise sense.

11. Expand $\dfrac{\sin 3z}{z^3}$ in Laurent series which converges for $0<|z|<R$ and find R for this function.

ANSWERS

1. Arc AB.
2. Along each path, the value of integral is

$$\frac{-1+5i}{6}.$$

3. (i) $\pi(\cos 1 + i \sin 1)$, (ii) $-\pi(\cos 1 - i \sin 1)$.
4. (i) $-\dfrac{3i\pi}{2}$, (ii) $\dfrac{3i\pi}{2}$, (iii) 0.
5. (i) $\sum_{k=0}^{\infty}(-1)^k\left(1-\frac{1}{2^{k+1}}\right)z^{2k}$,

(ii) $\sum_{k=0}^{\infty}\dfrac{(-1)^{k+1}z^{2k}}{2^{k+1}}+\sum_{k=0}^{\infty}\dfrac{(-1)^k}{z^{2k+2}}$,

(iii) $\sum_{k=0}^{\infty}\dfrac{(-1)^k(1-2^k)}{z^{2k+2}}$.

8. $-920 - 368\,i$
9. zero.
10. $\pi(\pi+4i)(1-i)/32\sqrt{2}$
11. $\dfrac{3}{z^2}+\dfrac{9}{z}+\dfrac{81}{40}z^2+\ldots, R\to\infty$.

Chapter 6

Singularities and Residues

Art 11. ZEROS AND SINGULARITIES

Let f(z) be analytic inside the circle C with its centre at $z = \alpha$ and radius r. Then by Taylor's theorem (10.49), we have

$$f(z) = a_0 + a_1(z-\alpha) + a_2(z-\alpha)^2 + \dots$$

$$= \sum_{s=0}^{\infty} a_s(z-\alpha)^s$$

for any z in this circle. Here $a_s = \dfrac{f^{(s)}(a)}{\angle s}$.

If $a_0 = a_1 = \dots = a_{k-1} = 0$, $a_k \neq 0$, then

$$f(z) = a_k(z-\alpha)^k + \dots$$

and f(z) is said to have a *zero of order* k at $z = \alpha$.

Now we suppose that f(z) is analytic in the circle $|z-\alpha| < \rho$ except at the point $z = \alpha$. If we surround the point $z = \alpha$ with two concentric circles of radii less than ρ, then by Laurent's theorem (10.55) we have the expansion

$$f(z) = a_0 + a_1(z-\alpha) + a_2(z-\alpha)^2 + \dots$$

$$+ \frac{b_1}{(z-\alpha)} + \frac{b_2}{(z-\alpha)^2} + \dots$$

The first part of this series containing the terms of the type $a_s(z-\alpha)^s$ where s is a non-negative number is analytic at $z = \alpha$. The remaining part of this series is not analytic and is known as the *principal* part of f(z) at $z = \alpha$. We now distinguish the following two cases:

i) When $b_k \neq 0, b_{k+1} = b_{k+2} = \ldots = 0$.

In this case the principal part of f(z) at $z = \alpha$ has a finite number of terms so that the expansion becomes

$$\frac{b_1}{(z-\alpha)} + \frac{b_2}{(z-\alpha)^2} + \ldots + \frac{b_k}{(z-\alpha)^k}$$

and f(z) is said to have a *pole of order* k. A pole is an example of *isolated singularity*. When k becomes unity, the singularity is known as the *simple pole*.

ii) When the number of terms in the principal part are infinite, *i.e.*

$$f(z) = \frac{b_1}{(z-\alpha)} + \frac{b_2}{(z-\alpha)^2} + \ldots,$$

the function f(z) is said to have an *essential singularity* at $z = \alpha$.

Examples (i) $f(z) = \sin^2 z$ has a zero of order two at $z = 0$.

(ii) $f(z) = \dfrac{1}{(z-2)(z-3)} = \dfrac{1}{z-3} - \dfrac{1}{z-2}$

has simple poles at $z = 2$ and $z = 3$.

(iii) $f(z) = e^{1/z} = 1 + \dfrac{1}{z} + \dfrac{1}{2z^2} + \ldots$

has an essential singularity at $z = 0$. This type of singularity is isolated for $z = 0$ and can be enclosed by the circle which has no other singularity.

Point at Infinity

To consider the behaviour of f(z) for large values $|z|$ we put $z = \dfrac{1}{\zeta}$ and consider the function $f\left(\dfrac{1}{\zeta}\right)$ for small values of $|\zeta|$. If $f\left(\dfrac{1}{\zeta}\right)$ has a zero or pole of order k at $\zeta = 0$, then f(z) is said to have a zero or pole of order k at the point at infinity.

Residues

Let f(z) be a function which has a singularity at $z = \alpha$.

Then by Laurent expansion (10.55)

$$f(z) = a_0 + a_1(z-\alpha) + a_2(z-\alpha)^2 + \ldots$$

$$+ \frac{b_1}{(z-\alpha)} + \frac{b_2}{(z-\alpha)^2} + \ldots$$

The coefficient b_1 of $\frac{1}{(z-\alpha)}$ is called the residue of f(z) at the singular point $z = \alpha$.

In case of simple pole at $z = \alpha$, f(z) has the expansion

$$f(z) = a_0 + a_1(z-\alpha) + a_2(z-\alpha)^2 + \ldots + \frac{b_1}{(z-\alpha)},$$

and so the residue of f(z) is

$$b_1 = \lim_{z \to \alpha} (z-\alpha)\, f(z). \qquad (11.1)$$

If $z = \alpha$ is a pole of order n, then

$$b_1 = \frac{1}{\angle n-1} \lim_{z \to \alpha} \frac{d^{n-1}}{dz^{n-1}} \{(z-\alpha)^n f(z)\}. \qquad (11.2)$$

Another Method

Since the residue at $z = \alpha$ is the coefficient of $\frac{1}{(z-\alpha)}$ in the Laurent expansion of f(z), the residue is the coefficient of $\frac{1}{t}$ in the expansion of $f(\alpha + t)$ as a power series where $|t|$ is sufficiently small.

Example: (i) Find the residue of $\frac{2z}{(z+4)(z-1)^2}$ at $z = -4$.

Solution: Here $f(z) = \frac{2z}{(z+4)(z-1)^2}$ and $z = -4$ is a pole of order one (simple pole). Hence, the residue of f(z) at $z = -4$ is

$$= \lim_{z \to -4} \left[(z+4)\frac{2z}{(z+4)(z-1)^2}\right]$$

$$= \lim_{z \to -4} \left[\frac{2z}{(z-1)^2}\right] = -\frac{8}{25.}$$

(ii) Find the residue of $f(z)=\dfrac{z^3}{(z-1)^4(z-2)(z-3)}$ at z = 1.

Solution: The function f(z) has a pole of order four at z = 1. Therefore, the residue of f(z) at z = 1 by (11.2) is

$$=\frac{1}{\angle 3}\lim_{z\to 1}\frac{d^3}{dz^3}\{(z-1)^4 f(z)\}$$

$$=\frac{1}{\angle 3}\lim_{z\to 1}\frac{d^3}{dz^3}\left\{\frac{z^3}{(z-2)(z-3)}\right\}$$

$$=\frac{1}{\angle 3}\lim_{z\to 1}\left\{\frac{48}{(z-2)^4}-\frac{162}{(z-3)^4}\right\}=\frac{101}{16}.$$

(iii) Find the residue of f(z) = cosec z at its singular points.

Solution: The singular points of $f(z)=\dfrac{1}{\sin z}$ are $z = n\pi$

where $n = 0, \pm 1, \pm 2, \ldots$

Now

$$f(z)=\frac{1}{\sin z}$$

$$=\frac{1}{\sin(n\pi+z-n\pi)}$$

$$=\frac{1}{\cos n\pi\,\sin(z-n\pi)}$$

$$=(-1)^n\left[\frac{1}{(z-n\pi)}-\frac{1}{\angle 3(z-n\pi)^3}+\ldots\right]$$

∴ Residue of f(z) at $z = n\pi$ is

$$=\text{coeff. of}\ \frac{1}{(z-n\pi)}\ \text{in the expansion of } f(z)$$

$$=(-1)^n.$$

(iv) What is the residue of f(z) = cot z at its singular points?

Solution: The singular points of $f(z)=\dfrac{\cos z}{\sin z}$ are $z = n\pi$

where $n = 0, \pm 1, \pm 2, \ldots..$

Thus the residue of f(z) at $z = n\pi$ is

$$= \lim_{z \to n\pi} (z - n\pi) \frac{\cos z}{\sin z}$$

$$= \lim_{z \to n\pi} \left[(z - n\pi) \frac{\cos z}{\cos n\pi \sin (z - n\pi)} \right]$$

$$= \sec n\pi \lim_{z \to n\pi} \left[\frac{(z - n\pi) \cos z}{(z - n\pi) - \frac{(z - n\pi)^3}{\angle 3} + \ldots} \right]$$

$$= \sec n\pi \lim_{z \to n\pi} \left[\frac{\cos z}{1 - \frac{(z - n\pi)^2}{\angle 3} + \frac{(z - n\pi)^4}{\angle 5} \ldots\ldots} \right]$$

$$= \sec n\pi \cos n\pi$$

$$= 1.$$

(v) Find the residue of $f(z) = \dfrac{z^3 - z^2 + 1}{z^3}$ at infinity.

Solution: We use the substitution $z = \dfrac{1}{\zeta}$,

whence $f\left(\dfrac{1}{\zeta}\right) = 1 - \zeta + \zeta^3$

Thus the residue of f(z) at $z = \infty$

$$= \text{ the residue of } f\left(\frac{1}{\zeta}\right) \text{ at } \zeta = 0$$

$$= \lim_{\zeta \to 0} (1 - \zeta + \zeta^3)$$

$$= 1.$$

We shall now state and prove an important theorem in the theory of residues known as the *Cauchy's residue theorem*. It states that:

If a function f(z) is analytic on and inside a closed contour C except at a finite number of poles within C, then

$$\int_C f(z)dz = 2\pi i \text{ (sum of residues of f(z) at its poles inside C).} \qquad (11.3)$$

Proof: Let $\alpha_1, \alpha_2, \ldots, \alpha_n$ be the poles of f(z) within C. We enclose α_s (s = 1, 2, ...,n) with small circles γ_s of radii ρ_s. Then the function f(z) is analytic within and on the boundary of the domain bounded by C and the circles γ_s.

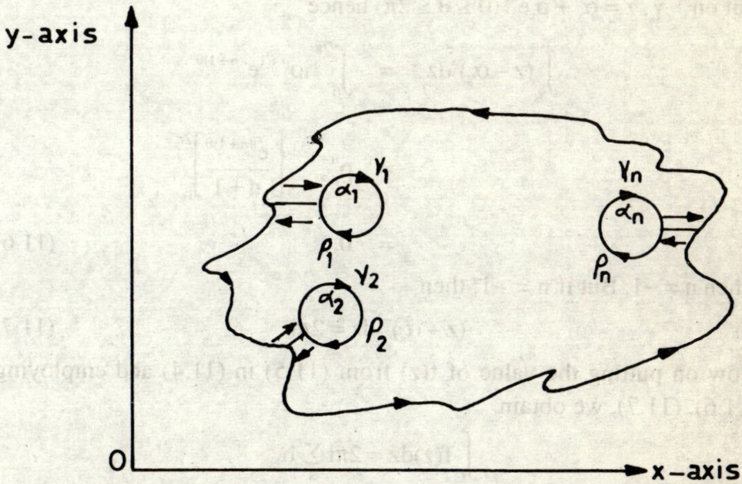

Fig. 63

Let λ_s be the curves connecting C with γ_s such that there are no singularities on λ_s (Fig. 63). Let Γ denote the composite boundary of C, γ_s, λ_s for all s. Then by *Cauchy Goursat* theorem (10.19)

$$\int_{\Gamma} f(z)dz = 0.$$

But

$$\int_{\Gamma} f(z)dz = \int_C f(z)\,dz + \sum_S \int_{\lambda_s} f(z)dz - \sum_s \int_{\lambda_s} f(z)\,dz$$

$$- \sum_S \int_{\gamma_s} f(z)dz$$

whence

$$\int_C f(z)dz = \sum_S \int_{\gamma_s} f(z)dz. \qquad (11.4)$$

We now suppose that α_s is a pole of order ρ. Then near $z = \alpha_s$

$$f(z) = a_{0s} + a_{1s}(z-\alpha_s) + \ldots$$

$$+\frac{b_{1s}}{(z-\alpha_s)} + \ldots + \frac{b_{ps}}{(z-\alpha_s)^p}. \tag{11.5}$$

But on γ_s, $z = \alpha_s + \rho_s e^{i\theta}$ $(0 \le \theta \le 2\pi)$ hence

$$\int_{\gamma_s} (z-\alpha_s)^n dz = \int_0^{2\pi} i\rho_s^{n+1} e^{i(n+1)\theta}$$

$$= \rho_s^{n+1} \left[\frac{e^{i(n+1)\theta}}{n+1}\right]_0^{2\pi}$$

$$= 0, \tag{11.6}$$

when $n \ne -1$. But if $n = -1$, then

$$(z-\alpha_s)^{-1} dz = 2\pi i. \tag{11.7}$$

Now on putting the value of f(z) from (11.5) in (11.4) and employing (11.6), (11.7), we obtain

$$\int_C f(z)dz = 2\pi i \sum_S b_{1s}$$

which proves the theorem.

Below we shall discuss the various applications of Cauchy-Residue theorem in evaluating the different types of definite integrals appearing in many physical problems of interest.

I. *Integration Around Unit Circle, Evaluation of Real Integrals*

Integrals of the type

$$I = \int_0^{2\pi} f(\cos\theta, \sin\theta)\, d\theta \tag{11.8}$$

where f is any rational function of $\cos\theta$ and $\sin\theta$ can be evaluated by taking the contour C as a unit circle $|z| = 1$.
For, any point on C is given by

$$z = e^{i\theta} \ (0 \le \theta \le 2\pi)$$

whence

$$\cos\theta = \frac{z + z^{-1}}{2}, \ \sin\theta = \frac{z - z^{-1}}{2i}, \ dz = ie^{i\theta} d\theta. \tag{11.9}$$

Using these values in (11.8) the definite integral (11.8) is replaced by a contour integral around C, *i.e.*,

$$I = \int_C f\left(\frac{z+z^{-1}}{2}, \frac{z-z^{-1}}{2i}\right)\frac{dz}{iz}$$

$$= \int_C F(z)\,dz \text{ say.}$$

Thus by residue theorem

$= 2\pi i$ (sum of the residues of F(z) at its poles within C). (11.10)

Example. Apply the calculus of residues to evaluate

$$I = \int_0^{\pi} \frac{a\,d\theta}{a^2+\cos^2\theta}, \quad \text{where a is a positive number.}$$

Solution: We have

$$I = \int_0^{\pi} \frac{a\,d\theta}{a^2+\cos^2\theta} = \int_0^{\pi} \frac{2a\,d\theta}{1+2a^2+\cos 2\theta}$$

$$= \int_0^{2\pi} \frac{a\,d\phi}{1+2a^2+\cos\phi}$$

by putting $\phi = 2\theta$.

If we use the substitution $z = e^{i\phi}$, then it becomes

$$I = \int_C \frac{-2ai\,dz}{z^2+2z(1+2a^2)+1}$$

where C denotes the unit circle $|z| = 1$.

The poles of integrand are given by

$$z^2 + 2z(1+2a^2) + 1 = 0$$

$$i.e., \quad z_1 = -(1+2a^2) + 2a\sqrt{(1+a^2)},$$

$$z_2 = -(1+2a^2) - 2a\sqrt{(1+a^2)}.$$

The pole z_2 lies outside the contour C because $|z_2| > 1$, a being a positive number. The only pole within C is z_1 whence the residue at $z = z_1$ is

$$= \lim_{z\to z_1}\left[(z-z_1)\frac{-2ai}{(z-z_1)(z-z_2)}\right]$$

$$\frac{-2ai}{z_1 - z_2} = -\frac{i}{2\sqrt{1+a^2}}.$$

Thus by Cauchy-Residue theorem

$$I = 2\pi i\,(-i/2\sqrt{1+a^2}) = \frac{\pi}{\sqrt{1+a^2}}.$$

II. *Infinite Integrals*

Theorem: If f(z) is a function such that (i) f(z) is analytic in the upper-half of z-plane [Im(z) ≥ 0] except for a finite number of poles,

(ii) f(z) has only simple poles on the real-axis,

(iii) | z f(z) | tends to the limit zero for all points z satisfying the condition

$$0 \le \arg \le \pi \quad \text{as} \quad z \to \infty,$$

then

Cauchy-principal value of $\int_{-\infty}^{\infty} f(x)dx$

= 2πi (sum of the residues of f(z) at its poles in the upper-half plane)

+πi (sum of the residues of f(z) at its simple poles on the real-axis).

(11.11)

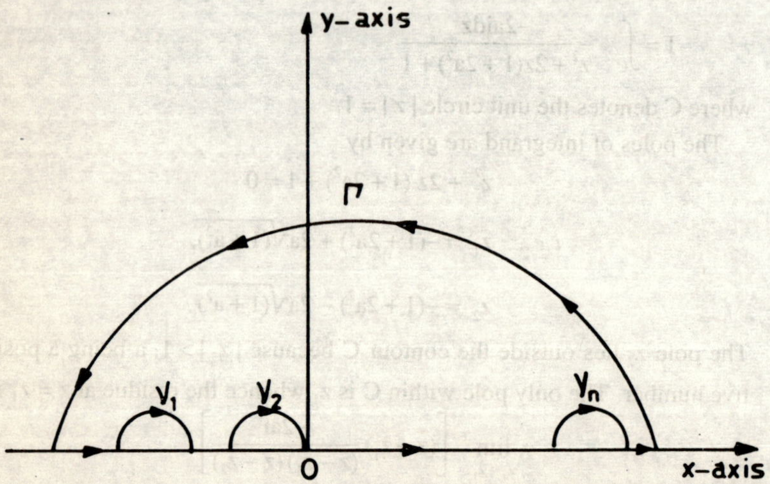

Fig. 64

Proof: Let $z=\alpha_s$ (s = 1, 2,...., n) be n-simple poles on the real-axis and let γ_s be n-semicircular indentations in the upper-half of z-plane with $z=\alpha_s$ as centres and ρ_s as radii (Fig. 64). Let C denote the composite contour consisting of the real-axis from –R to R (excluding the segments inside indentations), the semi-circles γ_s and the semi-circle Γ in the upper-half of z-plane having its centre at origin and radius R. Then by Cauchy-residue theorem (11.3)

$$\int_C f(z)dz = \int_\Gamma f(z)dz + \sum_{s=1}^{n} \int_{\gamma_s} f(z)\, dz + I_n$$

$$= 2\pi i \text{ (sum of the residues of f(z) at its poles inside C)} \qquad (11.12)$$

where

$$I_n = \int_{-R}^{\alpha_1-\rho_1} f(x)dx + \int_{\alpha_1+\rho_1}^{\alpha_2-\rho_2} f(x)dx + \ldots + \int_{\alpha_{n}+\rho_n}^{R} f(x)\, dx.$$

when $\rho_s \to 0$ for all s = 1, 2,..., n,

$$I_n \to \text{Cauchy-principal value of } \int_{-\infty}^{\infty} f(x)dx \text{ as } R \to \infty. \qquad (11.13)$$

We now consider

$$\int_{\gamma_s} f(z)\, dz.$$

On $\gamma_s, z-\alpha_s=\rho_s e^{i\theta}(0 \le \theta \le \pi)$.

Since f(z) has a simple pole at $z=\alpha_s$, we have from Laurent's expansion.

$$f(z) = \sum_{m=0}^{\infty} a_m(z-\alpha_s)^m + \frac{b_{1s}}{(z-\alpha_s)},$$

where b_{1s} is the residue of f(z) at $z=\alpha_s$.

Thus

$$\int_{\gamma_s} f(z)\, dz = \int_\pi^0 b_{1s}\, i\, d\theta + \sum_{m=0}^{\infty} \int_\pi^0 a_m\, i\, \rho_s^{m+1} e^{i(m+1)}\, d\theta.$$

As ρ_s tends to zero the last term of this equation becomes zero whence

$$\int_{\gamma_s} f(z)dz = -\pi i b_{1s}.$$

Thus

$$\sum_{s=1}^{n} \int_{\gamma_s} f(z)dz = -\sum_{s=1}^{n} \pi i\, b_{1s}$$

$= -\pi i$ (sum of the residues of f(z) at its simple poles on the real-axis)

(11.14)

Employing (11.13), (11.14) in (11.12) and letting $\rho_s \to 0$ for all s, we get

Cauchy-principal value of $\int_{-\infty}^{\infty} f(x)dx$

$= 2\pi i$ (sum of the residue of f(z) at its poles in the upper-half plane)

$+\pi i$ (sum of the residues of f(z) at its simple poles on the real-axis),

whence the theroem.

Particular Cases:

i) When there is no pole on real-axis, Cauchy-principal value of

$\int_{-\infty}^{\infty} f(x)dx$

$= 2\pi i$ (sum of the residues of f(z) at its pole in the upper-half plane).

(11.15)

ii) When there is no pole in the upper half of z-plane, Cauchy-principal value of $\int_{-\infty}^{\infty} f(x)\, dx$

$= \pi i$ (sum of the residues of f(z) at its simple poles on the real-axis).

(11.16)

Example (i) show that

$$\int_0^{\infty} \frac{dx}{1+x^2} = \frac{\pi}{2}.$$

Solution: We consider the integral $\int_C \frac{dz}{1+z^2}$ around the contour C (Fig. 65), nothing that there is no pole on the real-axis. Thus by (11.15)

$\int_{-\infty}^{\infty} \frac{dx}{1+x^2} = 2\pi i$ (sum of the residues of f(z) at its poles in the upper-half plane).

(11.17)

The poles of f(z) are given by

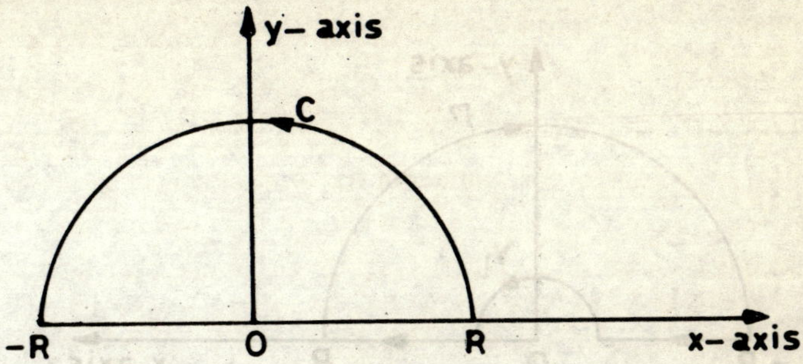

Fig. 65

$$1+z^2=0$$

i.e. $$z = \pm i.$$

Since $\rho > 1$, the only pole within C is at $z = i$. The residue at this pole is

$$= \lim_{z \to 1} \left[(z-i)\frac{1}{1+z^2} \right]$$

$$= \lim_{z \to 1} \left[\frac{1}{z+i} \right] = \frac{1}{2i}.$$

Hence from (11.16),

$$\int_{-\infty}^{\infty} \frac{dx}{1+x^2} = 2\pi i\left(\frac{1}{2i}\right) = \pi$$

i.e. $\int_0^{\infty} \frac{dx}{1+x^2} = \frac{\pi}{2}.$ Hence proved.

Example (ii) Evaluate

$$\int_0^{\infty} \frac{\sin mx}{x} dx$$

where m is a positive number.

Solution: Let $f(z) = \frac{e^{imz}}{z}$. We now consider the integral $\int_C f(z)\, dz$ along the contour C shown in Fig. 66. There is a simple pole at the origin, therefore, there is a semicircular indentation at 0, there being no pole within the contour C. Hence by (11.16)

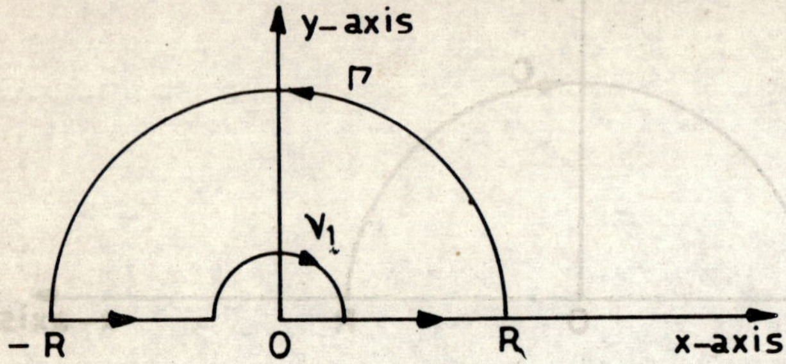

Fig. 66

Cauchy-principal value of $\int_{-\infty}^{\infty} \frac{e^{imx}}{x} dx$

$= \pi i$ (sum of the residues of f(z) at its simple poles on the real-axis). (11.18)

Residue of f(z) at z = 0 is

$$= \lim_{z \to 0} \left[z \frac{e^{imz}}{z}\right]$$

$= 1.$

Hence from (11.18),

Cauchy-principal value of $\int_{-\infty}^{\infty} \frac{e^{imx}}{x} dx = \pi i$ or on equating imaginary parts on both sides

$$\int_{-\infty}^{\infty} \frac{\sin mx}{x} dx = \pi i$$

$$i.e., \int_{0}^{\infty} \frac{\sin mx}{x} dx = \frac{\pi i}{2}.$$

Example (iii) Evaluate the Cauchy-principal value of $\int_{0}^{\infty} \frac{x^{2m}}{1 - x^{2n}} dx$ where $n > m \geq 0$, n and m being integers.

Solution: Here $f(z) = \frac{z^{2m}}{1 - z^{2n}}$. The poles of f(z) are simple poles at $z = e^{\frac{\pi i k}{n}}$

where $0 \leq k \leq 2n - 1$. There are two simple poles at $z = \pm 1$ arising due to $k = 0$, n. Of the remaining poles only those for which $1 \leq k \leq (n-1)$ lie in the upper-half of the plane. Let $\alpha = e^{(\pi i k/n)}$ for $0 \leq k \leq n$; then $\alpha^n = 1$.

Now residue of f(z) at $z = \alpha$ is

$$\lim_{z \to \alpha} \left[(z - \alpha) \quad \frac{z^{2m}}{1 - z^{2n}} \right]$$

$$= \alpha^{2m} \quad \lim_{z \to \alpha} \left[\frac{1}{-2nz^{2n-1}} \right] = -\frac{\alpha^{2m-n+1}}{2n}$$

If $\alpha = \pm 1$, then the residues of f(z) at $z = \pm 1$ are, respectively, $\frac{-1}{2n}$ and $\frac{1}{2n}$.

The sum of the residues in the upper-half of z-plane is

$$= -\frac{1}{2n} \sum_{k=1}^{n-1} \beta^k$$

$$= \frac{(\beta - \beta^n)}{2n(\beta - 1)}$$

where $\beta = e^{\frac{(2m+1)\pi i}{n}}$

But $\beta^n = e^{(2m+1)\pi i} = -1$,

whence the sum of residues in the upper-half of z-plane is

$$= \frac{\beta + 1}{2n(\beta - 1)} = \frac{1}{2n} \quad \frac{\beta^{1/2} + \beta^{-1/2}}{\beta^{1/2} - \beta^{-1/2}}$$

$$= \frac{1}{2ni} \cot\left(\frac{2m+1}{2n} \right)\pi.$$

Now from (11.11),

Cauchy-principal value of $\int_{-\infty}^{\infty} \frac{x^{2m}}{1 - x^{2n}} dx$

$$= 2\pi i \left(\frac{1}{2ni} \right) \cot\left(\frac{2m+1}{2n} \pi \right)$$

$$+ \pi i \left(-\frac{1}{2n} + \frac{1}{2n} \right)$$

$$=\frac{\pi}{n}\cot\left(\frac{2m+1}{2n}\right)\pi$$

i.e., Cauchy-principal value of $\int_0^\infty \frac{x^{2m}}{1-x^{2n}}dx$

$$=\frac{\pi}{2n}\cot\left(\frac{2m+1}{2n}\right)\pi.$$

III. *Jordan's Lemma*

Let C be a semicircle of radius R and centre at origin in the upper-half of z-plane and let f(z) be a function such that

(i) f(z) is analytic in the upper-half of z-plane except for a finite number of poles,

(ii) $|f(z)| \to 0$ as $R \to \infty$ for all values of z satisfying $0 \le \arg z \le \pi$,

then

$$\lim_{F \to \infty} \int e^{imz} f(z) = 0, \tag{11.19}$$

where $m > 0$.

Proof: We take the radius R so large as to include all poles of f(z) within C and none lying on its boundary.

From (ii) condition, we have for all points z on C

$$|f(z)| < \varepsilon \tag{11.20}$$

where ε is an arbitrary small positive number.

Now $\left|\int_C e^{imz} f(z)dz\right|$

$$\le \int_C |e^{imz}| \quad |f(z)| \quad |dz|$$

$$< \varepsilon \int_C |e^{imz}| \quad |dz|$$

by using (11.20).

If we use the substitution $z = R\,e^{i\theta}$, then it takes the form

$$\left|\int_C e^{imz} f(z)\,dz\right| < \varepsilon \int_0^\pi |e^{im(R\cos\theta + iR\sin\theta)}| \quad |Re^{i\theta} i d\theta|$$

$$= 2\varepsilon R \int_0^{\pi/2} e^{-mR\sin\theta} d\theta$$

$$= 2\varepsilon R \int_0^{\pi/2} e^{\frac{-2mR\theta}{\pi}} d\theta$$

where we have used the fact that $\frac{2}{\pi} \le \frac{\sin\theta}{\theta} \le 1$.

Hence $\left| \int_C e^{imz} f(z) dz \right| < \frac{\varepsilon\pi}{m}(1 - e^{-mR})$.

The right hand side of this inequality tends to the limit zero as R approaches infinity which proves the lemma.

Example. Show that if $a > 0$, $b > 0$,

$$\int_0^{\infty} \frac{\cos x}{(x^2 + a^2)(x^2 + b^2)} dx = \frac{\pi(ae^{-b} - be^{-a})}{ab(a^2 - b^2)}.$$

Solution: We consider the integral

$\int_c \frac{e^{iz}}{(z^2 + a^2)(z^2 + b^2)} dz$ where C is a closed contour consisting of a large circle γ of radius R and the real-axis from $-R$ to R. Then by Cauchy-Residue theorem

$$\int_\gamma \frac{e^{iz}}{(z^2 + a^2)(z^2 + b^2)} dz + \int_{-R}^{R} \frac{e^{ix}}{(x^2 + a^2)(x^2 + b^2)} dx$$

$= 2\pi i$ (sum of the residues of f(z) at its poles within C)

(11.21)

The poles of f(z) are given by

$$(z^2 + a^2)(z^2 + b^2) = 0$$

or $$z = \pm ia, \pm ib.$$

The only poles within C are $z = ia$, $z = ib$.

Residue of f(z) at $z = ia$ is

$$= \lim_{z \to ia} \left[(z - ia) \frac{e^{iz}}{(z^2 + a^2)(z^2 + b^2)} \right]$$

$$= -\frac{e^{-a}}{2ia(a^2 - b^2)}. \tag{11.22}$$

Similarly, residue of f(z) at $z = ib$ is

$$= -\frac{e^{-b}}{2ib(b^2 - a^2)} \tag{11.23}$$

Now we observe that $|f(z)| \to 0$ as $R \to \infty$.

Hence by Jordan's lemma (11.19)

$$\lim_{R \to \infty} \int_{\gamma} \frac{e^{iz}}{(z^2 + a^2)(z^2 + b^2)} dz = 0. \tag{11.24}$$

Letting $R \to \infty$ in Equation (11.21) and employing (11.22) to (11.24), we obtain

$$\int_{-\infty}^{\infty} \frac{e^{ix}}{(x^2 + a^2)(x^2 + b^2)} dx = \frac{\pi(ae^{-b} - be^{-a})}{ab(a^2 - b^2)}$$

which on equating real parts on both sides gives the required result.

IV. *Double Circle Contours*

Following example illustrates the application of double-circle contours in the calculus of residues.

Example. Evaluate

$$\int_0^{\infty} \frac{x^{a-1}}{1 + x} dx$$

where $0 < a < 1$

Solution: Let $f(z) = \dfrac{z^{a-1}}{1 + z}$. We now consider the integral $\int_C f(z)dz$ taken round the contour C shown in the Fig. 66.

The poles of f(z) are given by

$$z^{1-a}(1 + z) = 0$$

or $\quad z = 0, z = -1.$

The only pole lying inside C is the simple pole at $z = -1$.

The residue of f(z) at this pole is

$$\lim_{z \to e^{\pi i}} \left[(1 + z).\frac{z^{a-1}}{1 + z}\right]$$

$$= e^{(a-1)\pi i}$$

If r_1, r_2 denote the radii of circles γ_1, γ_2 respectively, then by Cauchy-Residue theorem

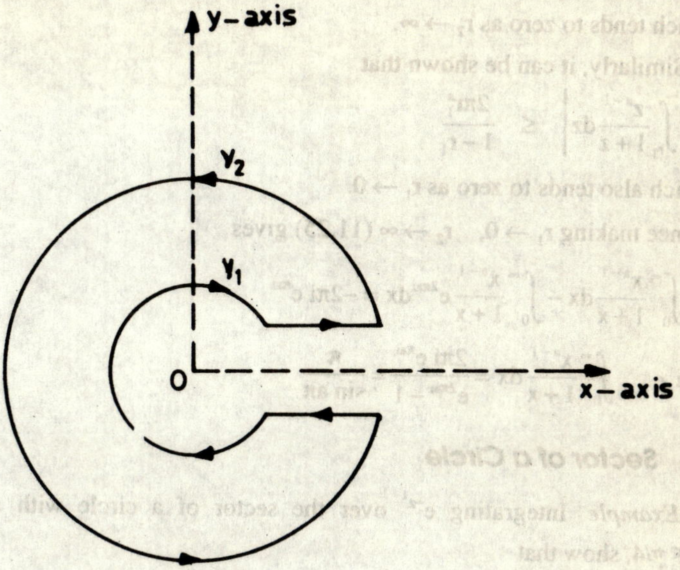

Fig. 67

$$\int_{r_1}^{r_2} f(x)dx + \int_{\gamma_2} f(z)\,dz - \int_{r_1}^{r_2} f(xe^{2\pi i})\,d(xe^{2\pi i})$$

$$-\int_{r_1} f(z)dz = 2\pi i\; e^{(a-1)\pi i}$$

i.e. $$\int_{r_1}^{r_2} \frac{x^{a-1}}{1+x}dx + \int_{\gamma_2}\frac{z^{a-1}}{1+x}dz - \int_{r_1}^{r_2}\frac{x^{a-1}}{1+x}\quad e^{2\pi ai}dx$$

$$-\int_{r_1}\frac{z^{a-1}}{1+z}dz = -2\pi i \quad e^{\pi ai}. \tag{11.25}$$

Now on r_2, $z = r_2e^{i\theta}(0 \le \theta \le 2\pi)$ whence

$$\left|\int_{r_2}\frac{z^{a-1}}{1+z}dz\right| \le \int_{r_2}\frac{|z|^{a-1}}{|1+z|}\ |dz|$$

$$\le \frac{2\pi r_2^{\,a}}{r_2-1}$$

which tends to zero as $r_2 \to \infty$.

Similarly, it can be shown that

$$\left|\int_{r_1} \frac{z^{a-1}}{1+z} dz\right| \le \frac{2\pi r_1^a}{1-r_1}$$

which also tends to zero as $r_1 \to 0$.

Hence making $r_1 \to 0, \quad r_2 \to \infty$ (11.25) gives

$$\int_0^\infty \frac{x^{a-1}}{1+x} dx - \int_0^\infty \frac{x^{a-1}}{1+x} e^{2\pi a i} dx = -2\pi i\, e^{\pi a i}$$

$$i.e., \quad \int_0^\infty \frac{x^{a-1}}{1+x} dx = \frac{2\pi i\, e^{\pi a i}}{e^{2\pi a i} - 1} = \frac{\pi}{\sin a\pi}$$

V. *Sector of a Circle*

Example. Integrating e^{-z^2} over the sector of a circle with angle $\alpha \le \pi/4$, show that

$$\int_0^\infty e^{-x^2 \cos 2\alpha} \cos(x^2 \sin 2\alpha)\, dx = \frac{\sqrt{\pi}}{2} \cos \alpha, \tag{11.26}$$

$$\int_0^\infty e^{-x^2 \cos 2\alpha} \sin(x^2 \sin 2\alpha) dx = \frac{\sqrt{\pi}}{2} \sin \alpha, \tag{11.27}$$

and deduce the result

$$\int_0^\infty \cos x^2\, dx = \int_0^\infty \sin x^2\, dx = \frac{\sqrt{\pi}}{2\sqrt{2}}. \tag{11.28}$$

Solution: We consider $\int_C e^{-z^2} dz$ around the contour C (Fig. 68).

Since f(z) is analytic within C, by Cauchy-residue theorem we have

$$\int_0^r e^{-x^2} dx + \int_\gamma e^{-z^2} dz + \int_r^0 e^{-(re^{i\alpha})^2} d(re^{i\alpha}) = 0.$$

as on OB, $z = re^{i\alpha}$. (11.29)

Now

$$\left|\int_\gamma e^{-z^2} dz\right| \le \int_\gamma |e^{-z^2}|\; |dz|$$

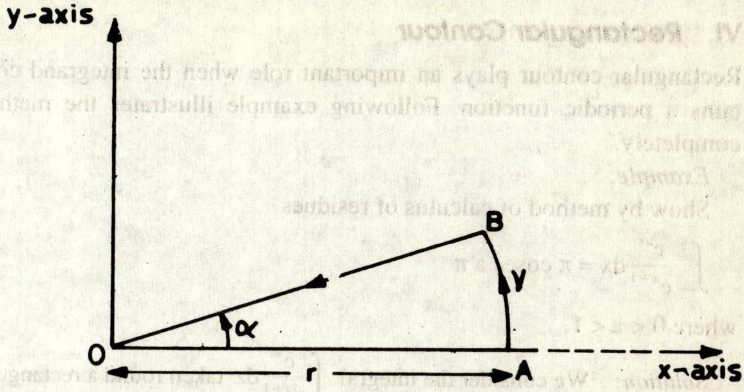

Fig. 68.

$$\leq \; r \int_0^{\infty} |e^{-r^2 e^{2i\theta}}| \; |id\theta|$$

$$< \frac{1}{2} r \int_{\beta}^{\pi/2} e^{-\frac{2r^2\phi}{\pi}} d\phi$$

by using the substitutions $\phi = \frac{\pi}{2} - 2\theta$, $\beta = \frac{\pi}{2} - 2\alpha$ and employing $\frac{2}{\pi} \leq \frac{\sin\theta}{\theta} \leq 1$

That is

$$\left| \int_{\gamma} e^{-z^2} dz \right| < \frac{\pi}{4r}\left(e^{-\frac{2r^2\beta}{\pi}} - e^{-r^2}\right)$$

which tends to the limit zero as r approaches the infinite values. Thus making $r \to \infty$ (11.29) becomes

$$\int_0^{\infty} e^{-x^2} dx - \int_0^{\infty} e^{-r^2} e^{2i\alpha} . e^{i\alpha} dr = 0$$

i.e., $$e^{-i\alpha} \int_0^{\infty} e^{-x^2} dx = \int_0^{\infty} e^{-r^2 \cos 2\alpha} . e^{-ir^2 \sin 2\alpha} dr$$

or

$$(\cos\alpha - i\sin\alpha)\frac{\sqrt{\pi}}{2} = \int_0^{\infty} e^{-x^2 \cos 2\alpha}[\cos(x^2 \sin 2\alpha) - i\sin(x^2 \sin 2\alpha)]\,dx$$

which on equating real and imaginary parts give (11.26) and (11.27).

If we put $\alpha = \pi/4$ in (11.26), (11.27), we obtain the result (11.28). Hence proved.

VI. Rectangular Contour

Rectangular contour plays an important role when the integrand contains a periodic function. Following example illustrates the method completely.

Example.

Show by method of calculus of residues

$$\int_{-\infty}^{\infty} \frac{e^{ax}}{e^{x+1}} dx = \pi \operatorname{cosec} a\pi$$

where $0 < a < 1$.

Solution: We consider the integral $\int_C \frac{e^{az}}{e^{z+1}} dz$ taken round a rectangular contour C formed by the lines $x = \pm r$, x-axis and $y = 2\pi$ (Fig. 69).

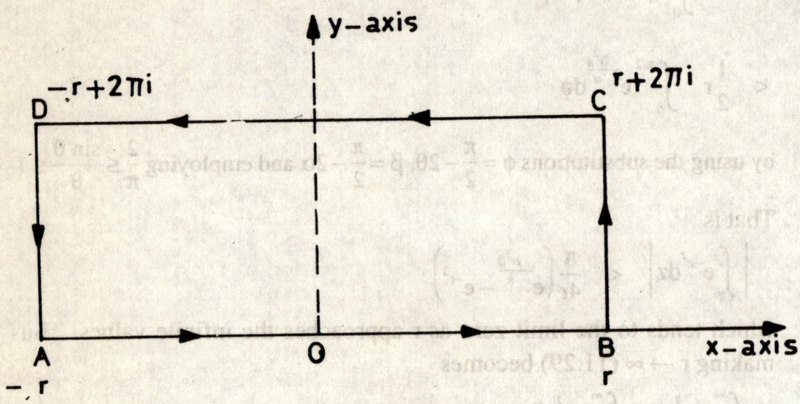

Fig. 69

The poles of f(z) are given by

$$1 + e^{z} = 0$$

i.e., $z = (2n+1)\pi i \quad (n = 0, 1, 2, \ldots)$.

The only pole within C is $z = \pi i$. The residue at this pole is $e^{-a\pi i}$

Hence by Cauchy-residue theorem

$$\int_{-r}^{r} f(x)dx + \int_{0}^{2\pi} f(r+iy)\, idy + \int_{r}^{-r} f(x+2\pi i)dx$$

$$+\int_0^{2\pi} f(-r+2\pi i)i\,dy = 2\pi i. - e^{a\pi i} \qquad (11.30)$$

Now

$$\left|\int_0^{2\pi} f(r+iy)i dy\right| \le \int_0^{2\pi} |f(r+iy)|\ |i|\ |dy|$$

$$\le \int_0^{2\pi} \frac{|e^{a(r+iy)}|}{|e^{r+iy}|-1} |dy|$$

$$\le \frac{2\pi e^{ar}}{e^r - 1}$$

which tends to zero as r approaches infinity.

Similarly, it can be shown that

$$\left|\int_0^{2\pi} f(-r+2\pi i)\, i dy\right| \to 0 \text{ as } r \to \infty$$

Hence, letting r tend to infinity (11.30) gives

$$\int_{-\infty}^{\infty} f(x)dx + \int_{\infty}^{-\infty} f(x+2\pi i)dx = -2\pi i e^{a\pi i}$$

from which we readily obtain

$$\int_{-\infty}^{\infty} \frac{e^{ax}}{e^x+1} dx = \pi \operatorname{cosec} a\pi.$$

Hence the result.

12. Meromorphic Function

Definition: A function whose only singularities in the finite part of the complex plane are poles is called a *meromorphic* function. A few examples of meromorphic function are tan z, cosec z, coth z, etc.

We shall now state and prove a theorem for such type of functions.

Theorem: If

i) f(z) is meromorphic inside a closed contour C;

ii) $f(z) \neq 0$ on C;

iii) N is the number of zeros and ρ is the number of poles of f(z) within C, where a pole (zero) of order m is counted m times, then

$$\frac{1}{2\pi i} \int_C \frac{f'(z)}{f(z)} dz = N - P. \qquad (12.1)$$

Proof: Let $z = \alpha_1$ be a zero of order m_1, then in the neighbourhood of $z = \alpha_1$,

$$f(z) = (z-\alpha_1)^{m_1} g(z)$$

where g(z) is analytic and non-zero. Thus

$$(f)'(z) = (z-\alpha_1)^{m_1} g'(z) + m_1(z-\alpha_1)^{m_1-1} g(z)$$

whence

$$\frac{f'(z)}{f(z)} = \frac{g'(z)}{g(z)} + \frac{m_1}{(z-\alpha_1)}.$$

This shows that the function $\frac{f'(z)}{f(z)}$ has a simple pole with residue m_1 at $z = \alpha_1$.

If $z = \beta_1$ is a pole of order n_1, then in the neighbourhood of $z = \beta_1$

$$f(z) = \frac{h(z)}{(z-\beta_1)^{n1}},$$

where h(z) is analytic. Hence

$$\frac{f'(z)}{f(z)} = \frac{h'(z)}{h(z)} - \frac{n_1}{(z-\beta_1)}$$

It means that $\frac{f'(z)}{f(z)}$ has a simple pole with residue $-n_1$ at $z = \beta_1$.

Extending these results we suppose that f(z) has zeros at $z = \alpha_1, \alpha_2, \ldots, \alpha_k$ of orders $m_1, m_2, \ldots, m_k$ respectively, and poles at $z = \beta_1, \beta_2, \ldots \beta_l$ of order $n_1, n_2, \ldots, n_l$ respectively. Then $\frac{f'(z)}{f(z)}$ has simple poles with residues $m_1, m_2, \ldots, m_k$ at $z = \alpha_1, \alpha_2, \ldots, \alpha_k$, respectively, and simple poles with residues $-n_1, -n_2, \ldots, -n_l$ at $z = \beta_1, \beta_2, \ldots, \beta_l$, respectively. Hence by Cauchy-Residue theorem

$$\int_C \frac{f'(z)}{f(z)} dz = 2\pi i \quad \left(\text{sum of the residues of } \frac{f'(z)}{f(z)} \text{ at its poles within C}\right).$$

$$= 2\pi i \left(\sum_{r=1}^{k} m_r - \sum_{r=1}^{l} n_r \right)$$

i.e. $$\frac{1}{2\pi i} \int_C \frac{f'(z)}{f(z)} dz = N - P.$$

which proves the theorem.

Corollary: If f(z) is analytic at all points interior to C then P = 0. So (12.1) simplifies to

$$\frac{1}{2\pi i}\int_C \frac{f'(z)}{f(z)}dz = N. \qquad (12.2)$$

But $\dfrac{f'(z)}{f(z)} = \dfrac{d}{dz}\{\log f(z)\}$ whence

$$\frac{1}{2\pi i}\int_C \frac{f'(z)}{f(z)}dz = \frac{1}{2\pi i} \text{ [variation of log f(z) around C]} \qquad (12.3)$$

To evaluate right hand side of this equation we have log f(z) = log | f(z) | + i arg f(z) where log | f(z) | being single valued returns to its original value on completing a circuit of contour C. Hence (12.3) and (12.2) combine to give

$$N = \frac{1}{2\pi} \text{ (variation in arg f(z) around C)}. \qquad (12.4)$$

13. Summation of Series by Contour Integration

Series can also be summed up with the help of contour integration. The following theorem is note worthy in this connection.

Theorem: If f(z) is a meromorphic function and C is a contour enclosing the zeros m, m + 1, ..., n of sin π z, then

$$\sum_{r=m}^{n} f(r) = \frac{1}{2\pi i}\int_C \pi \cot \pi z\, f(z)dz - S_1, \qquad (13.1)$$

$$\sum_{r=m}^{n} (-1)^r f(r) = \frac{1}{2\pi i}\int_C \pi \operatorname{cosec} \pi z\, f(z)dz - S_2, \qquad (13.2)$$

where S_1, S_2 are the sum of the residues of π cot πz f(z), π cosec πz f(z) at the poles of f(z) within C.

Proof: Trivially

$$\int_C \pi\cot \pi z\, f(z)dz = 2\pi i \text{ (sum of the residues at poles of integrand)}$$

$$= 2\pi i\left[S_1 + \sum_{r=m}^{n} f(r)\right]$$

since the residue of π cot πz f(z) at the pole z = r of π cot πz is f(r).

Similarly,

$$\int_C \pi \operatorname{cosec} \pi z\, f(z)dz = 2\pi i \left[S_2 + \sum_{r=m}^{n} (-1)^r f(r)\right]$$

Hence the *theorem*.

Example. Evaluate $\sum_{n=1}^{\infty} \frac{2x}{x^2 + n^2\pi^2}$,

and express $\frac{\sin x}{x}$ as an infinite product.

Solution: Let $f(z) = \frac{2x}{x^2 + \pi^2 z^2}$. By (13.1), we have

$$\sum_{n=-N}^{N} \frac{2x}{x^2 + n^2\pi^2} = \frac{1}{2\pi i}\int_C (\pi \cot \pi z) f(z)\, dz$$

–(sum of residues of integrand at poles of f(z) within C) (13.3)

where C is a contour which tends to infinity in all directions with N and which does not pass through any singularities of the integrand.

It can be easily shown that

$$\left| \frac{1}{2\pi i}\int_C (\pi \cot \pi z)\, f(z)dz \right| \to 0 \text{ as } N \to \infty.$$

Then, since the residue of integrand at each of the poles $z = \pm \frac{xi}{\pi}$ is the same

and since the series is absolutely convergent, we have

$$\lim_{N \to \infty} \sum_{n=-N}^{N} \frac{2x}{x^2 + n^2\pi^2} = -\left[2\pi \cot \left(\frac{\pi i x}{\pi}\right) \frac{2x}{\frac{2\pi^2 ix}{\pi}}\right]$$

i.e., $$\sum_{n=-\infty}^{\infty} \frac{2x}{x^2 + n^2\pi^2} = 2\coth x$$

i.e., $$\sum_{n=1}^{\infty} \frac{2x}{x^2 + n^2\pi^2} = \coth x - \frac{1}{x}$$

Integrating both sides with respect to x

$$\sum_{n=1}^{\infty} \log(x^2 + n^2\pi^2) = \log\left\{\frac{C \sinh x}{x}\right\}$$

where log C is the constant of integration.

Let $x \to 0$ then

$$\sum_{n=1}^{\infty} \log(n^2\pi^2) = \log C.$$

Hence,

$$\frac{\sinh x}{x} = \prod_{n=1}^{\infty} \left(1 + \frac{x^2}{n^2\pi^2}\right).$$

This product converges for all values of x whether real or complex and so in particular when $x = i\theta$. Thus

$$\frac{\sin \theta}{\theta} = \prod_{n=1}^{\infty} \left(1 - \frac{\theta^2}{n^2\pi^2}\right).$$

Hence the result.

EXERCISE

1. Evaluate the residues of $\dfrac{z^3}{(z-1)(z-2)(z-3)}$ at $z = 1, 2, 3$ and infinity and show that their sum is zero.

2. Find the residue of $\dfrac{1}{(z^2+a^2)^2}$ at $z = ia$.

3. Show that

$$\int_0^{2\pi} \frac{d\theta}{2+\cos\theta} = \frac{2\pi}{\sqrt{3}}.$$

4. By integrating $\dfrac{1}{(a+b\cos\theta)^2}$ round a suitable contour prove that

$$\int_0^{2\pi} \frac{d\theta}{(a+b\cos\theta)^2} = \frac{2\pi a}{(a^2-b^2)^{3/2}},$$

where $a > b > 0$.

5. Evaluate $\displaystyle\int_0^{\pi} \frac{1+2\cos\theta}{5+4\cos\theta}\,d\theta.$

6. Prove that $\displaystyle\int_{-\infty}^{\infty} \frac{dx}{(x^2+1)^3} = \frac{3\pi}{8}.$

7. Prove by contour integration that

$$\int_0^\infty \frac{\log(1+x^2)}{1+x^2}\,dx = \pi \log_e 2.$$

8. Evaluate the following if $a \geq 0, b \geq 0$

$$\int_0^\infty \frac{\cos 2ax - \cos 2bx}{x^2}\,dx$$

9. Prove that

$$\int_0^\infty \frac{\cos mx}{a^2+x^2}\,dx = \frac{\pi}{2a} e^{-ma},\ m \geq 0.$$

10. Show that

$$\int_0^\infty \frac{x^a}{(1+x^2)^2}\,dx = \frac{\pi(1-a)}{4\cos\frac{\pi a}{2}}$$

where $-1 < a < 3$.

11. Integrating $\dfrac{z}{1-ae^{-iz}}$ round the rectangle with vertices at $\pm\pi, \pm\pi \pm iR$, show that

$$\int_0^\pi \frac{ax \sin x}{1-2a\cos x+a^2}\,dx = \pi \log\left(\frac{1+a}{a}\right)$$

where $a \geq 1$.

12. By integrating $\dfrac{e^{iz}}{z+a}$ round the square with vertices at O, R, R + iR, iR, prove that

$$\int_0^\infty \frac{\sin x}{x^2+a^2}\,dx = \int_0^\infty \frac{xe^{-ax}}{1+x^2}\,dx\ (a > 0).$$

13. Show that

$$\int_0^\infty \cos x^2\,dx = \int_0^\infty \sin x^2\,dx = \frac{\sqrt{\pi}}{2\sqrt{2}}.$$

14. Prove that

$$\sum_{n=-\infty}^{\infty} (-1)^n \frac{\cos(n+\alpha)\theta}{n+\alpha} = \pi \operatorname{cosec} \pi\alpha$$

where $-\pi < \theta < \pi$.

15. Show that

$$\text{i)} \int_0^{\infty} \frac{dx}{1+x^4} = \frac{\pi}{2\sqrt{2}}, \quad \text{ii)} \int_0^{\infty} \frac{\sin x}{x}\, dx = \frac{\pi}{2}.$$

16. Evaluate $\int_C \frac{e^z}{\cos z}\, dz$ and $\int_C \frac{\cot z}{z}$ where C is the unit circle (counter-clockwise).

17. Find the value of $\int_0^{2\pi} \frac{d\theta}{1-2k\sin\theta+k^2}$ $(k^2 < 1)$

18. Evaluate.
$\int_0^{\infty} e^{-x^2} \cos 2bx\, dx$

ANSWERS

1. Residues at z = 1, 2, 3 and infinity are $\frac{1}{2}, -8, \frac{27}{2}$ and -6, respectively.

2. $\frac{1}{4ia^3}$.

5. 0.

8. $\pi(b-a)$.

16. Zero in both cases.

17. $\frac{2\pi}{1-k^2}$.

18. $e^{-b^2}\sqrt{\pi}$.